Mometrix
TEST PREPARATION

ORELA®

Middle Grades Mathematics Secrets Study Guide

DEAR FUTURE EXAM SUCCESS STORY

First of all, **THANK YOU** for purchasing Mometrix study materials!

Second, congratulations! You are one of the few determined test-takers who are committed to doing whatever it takes to excel on your exam. **You have come to the right place.** We developed these study materials with one goal in mind: to deliver you the information you need in a format that's concise and easy to use.

In addition to optimizing your guide for the content of the test, we've outlined our recommended steps for breaking down the preparation process into small, attainable goals so you can make sure you stay on track.

We've also analyzed the entire test-taking process, identifying the most common pitfalls and showing how you can overcome them and be ready for any curveball the test throws you.

Standardized testing is one of the biggest obstacles on your road to success, which only increases the importance of doing well in the high-pressure, high-stakes environment of test day. Your results on this test could have a significant impact on your future, and this guide provides the information and practical advice to help you achieve your full potential on test day.

Your success is our success

We would love to hear from you! If you would like to share the story of your exam success or if you have any questions or comments in regard to our products, please contact us at **800-673-8175** or **support@mometrix.com**.

Thanks again for your business and we wish you continued success!

Sincerely,
The Mometrix Test Preparation Team

Need more help? Check out our flashcards at:
http://MometrixFlashcards.com/ORELA

Written and edited by the Mometrix Exam Secrets Test Prep Team
Printed in the United States of America

TABLE OF CONTENTS

Introduction

Thank you for purchasing this resource! You have made the choice to prepare yourself for a test that could have a huge impact on your future, and this guide is designed to help you be fully ready for test day. Obviously, it's important to have a solid understanding of the test material, but you also need to be prepared for the unique environment and stressors of the test, so that you can perform to the best of your abilities.

For this purpose, the first section that appears in this guide is the **Secret Keys**. We've devoted countless hours to meticulously researching what works and what doesn't, and we've boiled down our findings to the five most impactful steps you can take to improve your performance on the test. We start at the beginning with study planning and move through the preparation process, all the way to the testing strategies that will help you get the most out of what you know when you're finally sitting in front of the test.

We recommend that you start preparing for your test as far in advance as possible. However, if you've bought this guide as a last-minute study resource and only have a few days before your test, we recommend that you skip over the first two Secret Keys since they address a long-term study plan.

If you struggle with **test anxiety**, we strongly encourage you to check out our recommendations for how you can overcome it. Test anxiety is a formidable foe, but it can be beaten, and we want to make sure you have the tools you need to defeat it.

Secret Key #1 – Plan Big, Study Small

There's a lot riding on your performance. If you want to ace this test, you're going to need to keep your skills sharp and the material fresh in your mind. You need a plan that lets you review everything you need to know while still fitting in your schedule. We'll break this strategy down into three categories.

Information Organization

Start with the information you already have: the official test outline. From this, you can make a complete list of all the concepts you need to cover before the test. Organize these concepts into groups that can be studied together, and create a list of any related vocabulary you need to learn so you can brush up on any difficult terms. You'll want to keep this vocabulary list handy once you actually start studying since you may need to add to it along the way.

Time Management

Once you have your set of study concepts, decide how to spread them out over the time you have left before the test. Break your study plan into small, clear goals so you have a manageable task for each day and know exactly what you're doing. Then just focus on one small step at a time. When you manage your time this way, you don't need to spend hours at a time studying. Studying a small block of content for a short period each day helps you retain information better and avoid stressing over how much you have left to do. You can relax knowing that you have a plan to cover everything in time. In order for this strategy to be effective though, you have to start studying early and stick to your schedule. Avoid the exhaustion and futility that comes from last-minute cramming!

Study Environment

The environment you study in has a big impact on your learning. Studying in a coffee shop, while probably more enjoyable, is not likely to be as fruitful as studying in a quiet room. It's important to keep distractions to a minimum. You're only planning to study for a short block of time, so make the most of it. Don't pause to check your phone or get up to find a snack. It's also important to **avoid multitasking**. Research has consistently shown that multitasking will make your studying dramatically less effective. Your study area should also be comfortable and well-lit so you don't have the distraction of straining your eyes or sitting on an uncomfortable chair.

The time of day you study is also important. You want to be rested and alert. Don't wait until just before bedtime. Study when you'll be most likely to comprehend and remember. Even better, if you know what time of day your test will be, set that time aside for study. That way your brain will be used to working on that subject at that specific time and you'll have a better chance of recalling information.

Finally, it can be helpful to team up with others who are studying for the same test. Your actual studying should be done in as isolated an environment as possible, but the work of organizing the information and setting up the study plan can be divided up. In between study sessions, you can discuss with your teammates the concepts that you're all studying and quiz each other on the details. Just be sure that your teammates are as serious about the test as you are. If you find that your study time is being replaced with social time, you might need to find a new team.

Secret Key #2 – Make Your Studying Count

You're devoting a lot of time and effort to preparing for this test, so you want to be absolutely certain it will pay off. This means doing more than just reading the content and hoping you can remember it on test day. It's important to make every minute of study count. There are two main areas you can focus on to make your studying count.

Retention

It doesn't matter how much time you study if you can't remember the material. You need to make sure you are retaining the concepts. To check your retention of the information you're learning, try recalling it at later times with minimal prompting. Try carrying around flashcards and glance at one or two from time to time or ask a friend who's also studying for the test to quiz you.

To enhance your retention, look for ways to put the information into practice so that you can apply it rather than simply recalling it. If you're using the information in practical ways, it will be much easier to remember. Similarly, it helps to solidify a concept in your mind if you're not only reading it to yourself but also explaining it to someone else. Ask a friend to let you teach them about a concept you're a little shaky on (or speak aloud to an imaginary audience if necessary). As you try to summarize, define, give examples, and answer your friend's questions, you'll understand the concepts better and they will stay with you longer. Finally, step back for a big picture view and ask yourself how each piece of information fits with the whole subject. When you link the different concepts together and see them working together as a whole, it's easier to remember the individual components.

Finally, practice showing your work on any multi-step problems, even if you're just studying. Writing out each step you take to solve a problem will help solidify the process in your mind, and you'll be more likely to remember it during the test.

Modality

Modality simply refers to the means or method by which you study. Choosing a study modality that fits your own individual learning style is crucial. No two people learn best in exactly the same way, so it's important to know your strengths and use them to your advantage.

For example, if you learn best by visualization, focus on visualizing a concept in your mind and draw an image or a diagram. Try color-coding your notes, illustrating them, or creating symbols that will trigger your mind to recall a learned concept. If you learn best by hearing or discussing information, find a study partner who learns the same way or read aloud to yourself. Think about how to put the information in your own words. Imagine that you are giving a lecture on the topic and record yourself so you can listen to it later.

For any learning style, flashcards can be helpful. Organize the information so you can take advantage of spare moments to review. Underline key words or phrases. Use different colors for different categories. Mnemonic devices (such as creating a short list in which every item starts with the same letter) can also help with retention. Find what works best for you and use it to store the information in your mind most effectively and easily.

Secret Key #3 – Practice the Right Way

Your success on test day depends not only on how many hours you put into preparing, but also on whether you prepared the right way. It's good to check along the way to see if your studying is paying off. One of the most effective ways to do this is by taking practice tests to evaluate your progress. Practice tests are useful because they show exactly where you need to improve. Every time you take a practice test, pay special attention to these three groups of questions:

- The questions you got wrong
- The questions you had to guess on, even if you guessed right
- The questions you found difficult or slow to work through

This will show you exactly what your weak areas are, and where you need to devote more study time. Ask yourself why each of these questions gave you trouble. Was it because you didn't understand the material? Was it because you didn't remember the vocabulary? Do you need more repetitions on this type of question to build speed and confidence? Dig into those questions and figure out how you can strengthen your weak areas as you go back to review the material.

Additionally, many practice tests have a section explaining the answer choices. It can be tempting to read the explanation and think that you now have a good understanding of the concept. However, an explanation likely only covers part of the question's broader context. Even if the explanation makes perfect sense, **go back and investigate** every concept related to the question until you're positive you have a thorough understanding.

As you go along, keep in mind that the practice test is just that: practice. Memorizing these questions and answers will not be very helpful on the actual test because it is unlikely to have any of the same exact questions. If you only know the right answers to the sample questions, you won't be prepared for the real thing. **Study the concepts** until you understand them fully, and then you'll be able to answer any question that shows up on the test.

It's important to wait on the practice tests until you're ready. If you take a test on your first day of study, you may be overwhelmed by the amount of material covered and how much you need to learn. Work up to it gradually.

On test day, you'll need to be prepared for answering questions, managing your time, and using the test-taking strategies you've learned. It's a lot to balance, like a mental marathon that will have a big impact on your future. Like training for a marathon, you'll need to start slowly and work your way up. When test day arrives, you'll be ready.

Start with the strategies you've read in the first two Secret Keys—plan your course and study in the way that works best for you. If you have time, consider using multiple study resources to get different approaches to the same concepts. It can be helpful to see difficult concepts from more than one angle. Then find a good source for practice tests. Many times, the test website will suggest potential study resources or provide sample tests.

Practice Test Strategy

If you're able to find at least three practice tests, we recommend this strategy:

Untimed and Open-Book Practice

Take the first test with no time constraints and with your notes and study guide handy. Take your time and focus on applying the strategies you've learned.

Timed and Open-Book Practice

Take the second practice test open-book as well, but set a timer and practice pacing yourself to finish in time.

Timed and Closed-Book Practice

Take any other practice tests as if it were test day. Set a timer and put away your study materials. Sit at a table or desk in a quiet room, imagine yourself at the testing center, and answer questions as quickly and accurately as possible.

Keep repeating timed and closed-book tests on a regular basis until you run out of practice tests or it's time for the actual test. Your mind will be ready for the schedule and stress of test day, and you'll be able to focus on recalling the material you've learned.

Secret Key #4 – Pace Yourself

Once you're fully prepared for the material on the test, your biggest challenge on test day will be managing your time. Just knowing that the clock is ticking can make you panic even if you have plenty of time left. Work on pacing yourself so you can build confidence against the time constraints of the exam. Pacing is a difficult skill to master, especially in a high-pressure environment, so **practice is vital**.

Set time expectations for your pace based on how much time is available. For example, if a section has 60 questions and the time limit is 30 minutes, you know you have to average 30 seconds or less per question in order to answer them all. Although 30 seconds is the hard limit, set 25 seconds per question as your goal, so you reserve extra time to spend on harder questions. When you budget extra time for the harder questions, you no longer have any reason to stress when those questions take longer to answer.

Don't let this time expectation distract you from working through the test at a calm, steady pace, but keep it in mind so you don't spend too much time on any one question. Recognize that taking extra time on one question you don't understand may keep you from answering two that you do understand later in the test. If your time limit for a question is up and you're still not sure of the answer, mark it and move on, and come back to it later if the time and the test format allow. If the testing format doesn't allow you to return to earlier questions, just make an educated guess; then put it out of your mind and move on.

On the easier questions, be careful not to rush. It may seem wise to hurry through them so you have more time for the challenging ones, but it's not worth missing one if you know the concept and just didn't take the time to read the question fully. Work efficiently but make sure you understand the question and have looked at all of the answer choices, since more than one may seem right at first.

Even if you're paying attention to the time, you may find yourself a little behind at some point. You should speed up to get back on track, but do so wisely. Don't panic; just take a few seconds less on each question until you're caught up. Don't guess without thinking, but do look through the answer choices and eliminate any you know are wrong. If you can get down to two choices, it is often worthwhile to guess from those. Once you've chosen an answer, move on and don't dwell on any that you skipped or had to hurry through. If a question was taking too long, chances are it was one of the harder ones, so you weren't as likely to get it right anyway.

On the other hand, if you find yourself getting ahead of schedule, it may be beneficial to slow down a little. The more quickly you work, the more likely you are to make a careless mistake that will affect your score. You've budgeted time for each question, so don't be afraid to spend that time. Practice an efficient but careful pace to get the most out of the time you have.

Secret Key #5 – Have a Plan for Guessing

When you're taking the test, you may find yourself stuck on a question. Some of the answer choices seem better than others, but you don't see the one answer choice that is obviously correct. What do you do?

The scenario described above is very common, yet most test takers have not effectively prepared for it. Developing and practicing a plan for guessing may be one of the single most effective uses of your time as you get ready for the exam.

In developing your plan for guessing, there are three questions to address:

- When should you start the guessing process?
- How should you narrow down the choices?
- Which answer should you choose?

When to Start the Guessing Process

Unless your plan for guessing is to select C every time (which, despite its merits, is not what we recommend), you need to leave yourself enough time to apply your answer elimination strategies. Since you have a limited amount of time for each question, that means that if you're going to give yourself the best shot at guessing correctly, you have to decide quickly whether or not you will guess.

Of course, the best-case scenario is that you don't have to guess at all, so first, see if you can answer the question based on your knowledge of the subject and basic reasoning skills. Focus on the key words in the question and try to jog your memory of related topics. Give yourself a chance to bring the knowledge to mind, but once you realize that you don't have (or you can't access) the knowledge you need to answer the question, it's time to start the guessing process.

It's almost always better to start the guessing process too early than too late. It only takes a few seconds to remember something and answer the question from knowledge. Carefully eliminating wrong answer choices takes longer. Plus, going through the process of eliminating answer choices can actually help jog your memory.

Summary: Start the guessing process as soon as you decide that you can't answer the question based on your knowledge.

How to Narrow Down the Choices

The next chapter in this book (**Test-Taking Strategies**) includes a wide range of strategies for how to approach questions and how to look for answer choices to eliminate. You will definitely want to read those carefully, practice them, and figure out which ones work best for you. Here though, we're going to address a mindset rather than a particular strategy.

Your odds of guessing an answer correctly depend on how many options you are choosing from.

Number of options left	5	4	3	2	1
Odds of guessing correctly	20%	25%	33%	50%	100%

You can see from this chart just how valuable it is to be able to eliminate incorrect answers and make an educated guess, but there are two things that many test takers do that cause them to miss out on the benefits of guessing:

- Accidentally eliminating the correct answer
- Selecting an answer based on an impression

We'll look at the first one here, and the second one in the next section.

To avoid accidentally eliminating the correct answer, we recommend a thought exercise called **the $5 challenge**. In this challenge, you only eliminate an answer choice from contention if you are willing to bet $5 on it being wrong. Why $5? Five dollars is a small but not insignificant amount of money. It's an amount you could afford to lose but wouldn't want to throw away. And while losing $5 once might not hurt too much, doing it twenty times will set you back $100. In the same way, each small decision you make—eliminating a choice here, guessing on a question there—won't by itself impact your score very much, but when you put them all together, they can make a big difference. By holding each answer choice elimination decision to a higher standard, you can reduce the risk of accidentally eliminating the correct answer.

The $5 challenge can also be applied in a positive sense: If you are willing to bet $5 that an answer choice *is* correct, go ahead and mark it as correct.

Summary: Only eliminate an answer choice if you are willing to bet $5 that it is wrong.

Which Answer to Choose

You're taking the test. You've run into a hard question and decided you'll have to guess. You've eliminated all the answer choices you're willing to bet $5 on. Now you have to pick an answer. Why do we even need to talk about this? Why can't you just pick whichever one you feel like when the time comes?

The answer to these questions is that if you don't come into the test with a plan, you'll rely on your impression to select an answer choice, and if you do that, you risk falling into a trap. The test writers know that everyone who takes their test will be guessing on some of the questions, so they intentionally write wrong answer choices to seem plausible. You still have to pick an answer though, and if the wrong answer choices are designed to look right, how can you ever be sure that you're not falling for their trap? The best solution we've found to this dilemma is to take the decision out of your hands entirely. Here is the process we recommend:

Once you've eliminated any choices that you are confident (willing to bet $5) are wrong, select the first remaining choice as your answer.

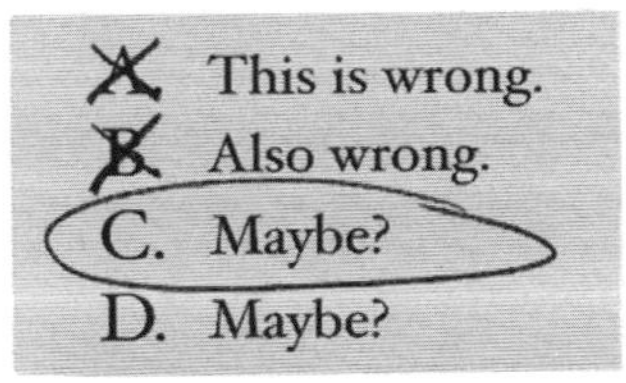

Whether you choose to select the first remaining choice, the second, or the last, the important thing is that you use some preselected standard. Using this approach guarantees that you will not be enticed into selecting an answer choice that looks right, because you are not basing your decision on how the answer choices look.

This is not meant to make you question your knowledge. Instead, it is to help you recognize the difference between your knowledge and your impressions. There's a huge difference between thinking an answer is right because of what you know, and thinking an answer is right because it looks or sounds like it should be right.

Summary: To ensure that your selection is appropriately random, make a predetermined selection from among all answer choices you have not eliminated.

Test-Taking Strategies

This section contains a list of test-taking strategies that you may find helpful as you work through the test. By taking what you know and applying logical thought, you can maximize your chances of answering any question correctly!

It is very important to realize that every question is different and every person is different: no single strategy will work on every question, and no single strategy will work for every person. That's why we've included all of them here, so you can try them out and determine which ones work best for different types of questions and which ones work best for you.

Question Strategies

✓ Read Carefully

Read the question and the answer choices carefully. Don't miss the question because you misread the terms. You have plenty of time to read each question thoroughly and make sure you understand what is being asked. Yet a happy medium must be attained, so don't waste too much time. You must read carefully and efficiently.

✓ Contextual Clues

Look for contextual clues. If the question includes a word you are not familiar with, look at the immediate context for some indication of what the word might mean. Contextual clues can often give you all the information you need to decipher the meaning of an unfamiliar word. Even if you can't determine the meaning, you may be able to narrow down the possibilities enough to make a solid guess at the answer to the question.

✓ Prefixes

If you're having trouble with a word in the question or answer choices, try dissecting it. Take advantage of every clue that the word might include. Prefixes can be a huge help. Usually, they allow you to determine a basic meaning. *Pre-* means before, *post-* means after, *pro-* is positive, *de-* is negative. From prefixes, you can get an idea of the general meaning of the word and try to put it into context.

✓ Hedge Words

Watch out for critical hedge words, such as *likely, may, can, often, almost, mostly, usually, generally, rarely,* and *sometimes.* Question writers insert these hedge phrases to cover every possibility. Often an answer choice will be wrong simply because it leaves no room for exception. Be on guard for answer choices that have definitive words such as *exactly* and *always.*

✓ Switchback Words

Stay alert for *switchbacks.* These are the words and phrases frequently used to alert you to shifts in thought. The most common switchback words are *but, although,* and *however.* Others include *nevertheless, on the other hand, even though, while, in spite of, despite,* and *regardless of.* Switchback words are important to catch because they can change the direction of the question or an answer choice.

⊘ Face Value

When in doubt, use common sense. Accept the situation in the problem at face value. Don't read too much into it. These problems will not require you to make wild assumptions. If you have to go beyond creativity and warp time or space in order to have an answer choice fit the question, then you should move on and consider the other answer choices. These are normal problems rooted in reality. The applicable relationship or explanation may not be readily apparent, but it is there for you to figure out. Use your common sense to interpret anything that isn't clear.

Answer Choice Strategies

⊘ Answer Selection

The most thorough way to pick an answer choice is to identify and eliminate wrong answers until only one is left, then confirm it is the correct answer. Sometimes an answer choice may immediately seem right, but be careful. The test writers will usually put more than one reasonable answer choice on each question, so take a second to read all of them and make sure that the other choices are not equally obvious. As long as you have time left, it is better to read every answer choice than to pick the first one that looks right without checking the others.

⊘ Answer Choice Families

An answer choice family consists of two (in rare cases, three) answer choices that are very similar in construction and cannot all be true at the same time. If you see two answer choices that are direct opposites or parallels, one of them is usually the correct answer. For instance, if one answer choice says that quantity x increases and another either says that quantity x decreases (opposite) or says that quantity y increases (parallel), then those answer choices would fall into the same family. An answer choice that doesn't match the construction of the answer choice family is more likely to be incorrect. Most questions will not have answer choice families, but when they do appear, you should be prepared to recognize them.

⊘ Eliminate Answers

Eliminate answer choices as soon as you realize they are wrong, but make sure you consider all possibilities. If you are eliminating answer choices and realize that the last one you are left with is also wrong, don't panic. Start over and consider each choice again. There may be something you missed the first time that you will realize on the second pass.

⊘ Avoid Fact Traps

Don't be distracted by an answer choice that is factually true but doesn't answer the question. You are looking for the choice that answers the question. Stay focused on what the question is asking for so you don't accidentally pick an answer that is true but incorrect. Always go back to the question and make sure the answer choice you've selected actually answers the question and is not merely a true statement.

⊘ Extreme Statements

In general, you should avoid answers that put forth extreme actions as standard practice or proclaim controversial ideas as established fact. An answer choice that states the "process should be used in certain situations, if..." is much more likely to be correct than one that states the "process should be discontinued completely." The first is a calm rational statement and doesn't even make a definitive, uncompromising stance, using a hedge word *if* to provide wiggle room, whereas the second choice is far more extreme.

✓ Benchmark

As you read through the answer choices and you come across one that seems to answer the question well, mentally select that answer choice. This is not your final answer, but it's the one that will help you evaluate the other answer choices. The one that you selected is your benchmark or standard for judging each of the other answer choices. Every other answer choice must be compared to your benchmark. That choice is correct until proven otherwise by another answer choice beating it. If you find a better answer, then that one becomes your new benchmark. Once you've decided that no other choice answers the question as well as your benchmark, you have your final answer.

✓ Predict the Answer

Before you even start looking at the answer choices, it is often best to try to predict the answer. When you come up with the answer on your own, it is easier to avoid distractions and traps because you will know exactly what to look for. The right answer choice is unlikely to be word-for-word what you came up with, but it should be a close match. Even if you are confident that you have the right answer, you should still take the time to read each option before moving on.

General Strategies

✓ Tough Questions

If you are stumped on a problem or it appears too hard or too difficult, don't waste time. Move on! Remember though, if you can quickly check for obviously incorrect answer choices, your chances of guessing correctly are greatly improved. Before you completely give up, at least try to knock out a couple of possible answers. Eliminate what you can and then guess at the remaining answer choices before moving on.

✓ Check Your Work

Since you will probably not know every term listed and the answer to every question, it is important that you get credit for the ones that you do know. Don't miss any questions through careless mistakes. If at all possible, try to take a second to look back over your answer selection and make sure you've selected the correct answer choice and haven't made a costly careless mistake (such as marking an answer choice that you didn't mean to mark). This quick double check should more than pay for itself in caught mistakes for the time it costs.

✓ Pace Yourself

It's easy to be overwhelmed when you're looking at a page full of questions; your mind is confused and full of random thoughts, and the clock is ticking down faster than you would like. Calm down and maintain the pace that you have set for yourself. Especially as you get down to the last few minutes of the test, don't let the small numbers on the clock make you panic. As long as you are on track by monitoring your pace, you are guaranteed to have time for each question.

✓ Don't Rush

It is very easy to make errors when you are in a hurry. Maintaining a fast pace in answering questions is pointless if it makes you miss questions that you would have gotten right otherwise. Test writers like to include distracting information and wrong answers that seem right. Taking a little extra time to avoid careless mistakes can make all the difference in your test score. Find a pace that allows you to be confident in the answers that you select.

✓ Keep Moving

Panicking will not help you pass the test, so do your best to stay calm and keep moving. Taking deep breaths and going through the answer elimination steps you practiced can help to break through a stress barrier and keep your pace.

Final Notes

The combination of a solid foundation of content knowledge and the confidence that comes from practicing your plan for applying that knowledge is the key to maximizing your performance on test day. As your foundation of content knowledge is built up and strengthened, you'll find that the strategies included in this chapter become more and more effective in helping you quickly sift through the distractions and traps of the test to isolate the correct answer.

Now that you're preparing to move forward into the test content chapters of this book, be sure to keep your goal in mind. As you read, think about how you will be able to apply this information on the test. If you've already seen sample questions for the test and you have an idea of the question format and style, try to come up with questions of your own that you can answer based on what you're reading. This will give you valuable practice applying your knowledge in the same ways you can expect to on test day.

Good luck and good studying!

Number Sense and Operations

Number Basics

Classifications of Numbers

Numbers are the basic building blocks of mathematics. Specific features of numbers are identified by the following terms:

Integer – any positive or negative whole number, including zero. Integers do not include fractions $\left(\frac{1}{3}\right)$, decimals (0.56), or mixed numbers $\left(7\frac{3}{4}\right)$.

Prime number – any whole number greater than 1 that has only two factors, itself and 1; that is, a number that can be divided evenly only by 1 and itself.

Composite number – any whole number greater than 1 that has more than two different factors; in other words, any whole number that is not a prime number. For example: The composite number 8 has the factors of 1, 2, 4, and 8.

Even number – any integer that can be divided by 2 without leaving a remainder. For example: 2, 4, 6, 8, and so on.

Odd number – any integer that cannot be divided evenly by 2. For example: 3, 5, 7, 9, and so on.

Decimal number – any number that uses a decimal point to show the part of the number that is less than one. Example: 1.234.

Decimal point – a symbol used to separate the ones place from the tenths place in decimals or dollars from cents in currency.

Decimal place – the position of a number to the right of the decimal point. In the decimal 0.123, the 1 is in the first place to the right of the decimal point, indicating tenths; the 2 is in the second place, indicating hundredths; and the 3 is in the third place, indicating thousandths.

The **decimal**, or base 10, system is a number system that uses ten different digits (0, 1, 2, 3, 4, 5, 6, 7, 8, 9). An example of a number system that uses something other than ten digits is the **binary**, or base 2, number system, used by computers, which uses only the numbers 0 and 1. It is thought that the decimal system originated because people had only their 10 fingers for counting.

Rational numbers include all integers, decimals, and fractions. Any terminating or repeating decimal number is a rational number.

Irrational numbers cannot be written as fractions or decimals because the number of decimal places is infinite and there is no recurring pattern of digits within the number. For example, pi (π) begins with 3.141592 and continues without terminating or repeating, so pi is an irrational number.

Real numbers are the set of all rational and irrational numbers.

Review Video: Classification of Numbers
Visit mometrix.com/academy and enter code: 461071

Review Video: Prime and Composite Numbers
Visit mometrix.com/academy and enter code: 565581

NUMBERS IN WORD FORM AND PLACE VALUE

When writing numbers out in word form or translating word form to numbers, it is essential to understand how a place value system works. In the decimal or base-10 system, each digit of a number represents how many of the corresponding place value—a specific factor of 10—are contained in the number being represented. To make reading numbers easier, every three digits to the left of the decimal place is preceded by a comma. The following table demonstrates some of the place values:

Power of 10	10^3	10^2	10^1	10^0	10^{-1}	10^{-2}	10^{-3}
Value	1,000	100	10	1	0.1	0.01	0.001
Place	thousands	hundreds	tens	ones	tenths	hundredths	thousandths

For example, consider the number 4,546.09, which can be separated into each place value like this:

4: thousands
5: hundreds
4: tens
6: ones
0: tenths
9: hundredths

This number in word form would be *four thousand five hundred forty-six and nine hundredths.*

Review Video: Place Value
Visit mometrix.com/academy and enter code: 205433

Number Lines

A number line is a graph to see the distance between numbers. Basically, this graph shows the relationship between numbers. So a number line may have a point for zero and may show negative numbers on the left side of the line. Any positive numbers are placed on the right side of the line. For example, consider the points labeled on the following number line:

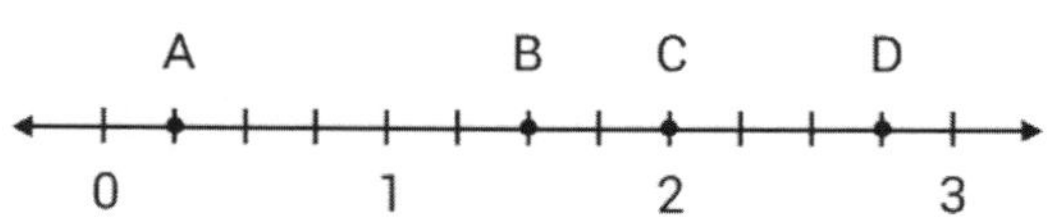

We can use the dashed lines on the number line to identify each point. Each dashed line between two whole numbers is $\frac{1}{4}$. The line halfway between two numbers is $\frac{1}{2}$.

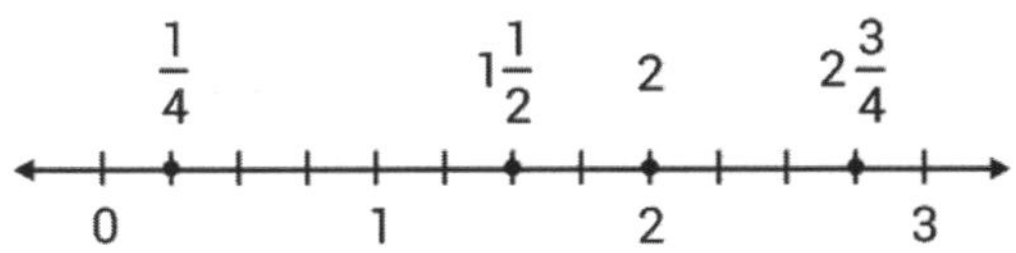

Review Video: The Number Line
Visit mometrix.com/academy and enter code: 816439

Comparing Numbers

Inequality Notation

The symbols $<$ and $>$ mean "is less than" and "is greater than," respectively. For instance, $3 < 5$ means "3 is less than 5," and $7 > 4$ means "7 is greater than 4." Statements like $3 < 5$ and $7 > 4$ are **inequalities**, and the symbols $<$ and $>$ are **inequality symbols**.

Whole Numbers and Decimal Numbers

To compare whole or decimal numbers, we look at the most significant place (the leftmost digit) at which they differ. The number with the larger digit in that place is larger. For instance, $0.3\underline{8}74$ and $0.3\underline{9}$ differ in the hundredths place (underlined). Since 8 is smaller than 9, we see $0.3874 < 0.39$. This is clearer if we make the decimals equal in length by writing extra zeroes: $0.3874 < 0.3900$. Similarly, $2\underline{3}.984 < 2\underline{5}.112$ because 3 is smaller than 5, or 23 is smaller than 25.

Fractions

If fractions have the same denominator, the fraction with the larger numerator is larger. For instance, $\frac{2}{7} < \frac{5}{7}$ since $2 < 5$. We compare fractions with different denominators by finding a common denominator. When comparing the fractions with a common denominator we only compare the numerator, so as a shortcut, we can multiply each numerator by the denominator of the other fraction. The numerator that produces the larger product belongs to the larger fraction. For example, to compare $\frac{7}{8}$ and $\frac{5}{6}$, we note that $7 \cdot 6 = 42$ is larger than $5 \cdot 8 = 40$. Since the numerator 7 produces the larger product, we see $\frac{7}{8} > \frac{5}{6}$. We can also compare fractions by converting them to decimals. For instance, since $\frac{3}{4} = 0.75$ and $\frac{4}{5} = 0.8$ and $0.75 < 0.8$, we conclude $\frac{3}{4} < \frac{4}{5}$.

Mixed Numbers

To compare mixed numbers we compare their whole number parts. If those are equal, then we compare their fractional parts. For instance, $5\frac{3}{8} > 4\frac{7}{8}$ because $5 > 4$, but $3\frac{5}{9} < 3\frac{8}{9}$ because $\frac{5}{9} < \frac{8}{9}$.

Square Roots

To compare square roots, we convert them to decimals, usually with a calculator. To compare two square roots, we compare their radicands. For instance, $\sqrt{11} < \sqrt{14}$ because $11 < 14$.

NEGATIVE NUMBERS

A negative number is always less than a positive number. Two negative numbers compare in the reverse order of their opposites. For instance, $-6 < -2$ (that is, -6 is smaller, more negative, than -2) because $6 > 2$.

Absolute Value

A precursor to working with negative numbers is understanding what **absolute values** are. A number's absolute value is simply the distance away from zero a number is on the number line. The absolute value of a number is always positive and is written $|x|$. For example, the absolute value of 3, written as $|3|$, is 3 because the distance between 0 and 3 on a number line is three units. Likewise, the absolute value of -3, written as $|-3|$, is 3 because the distance between 0 and -3 on a number line is three units. So $|3| = |-3|$.

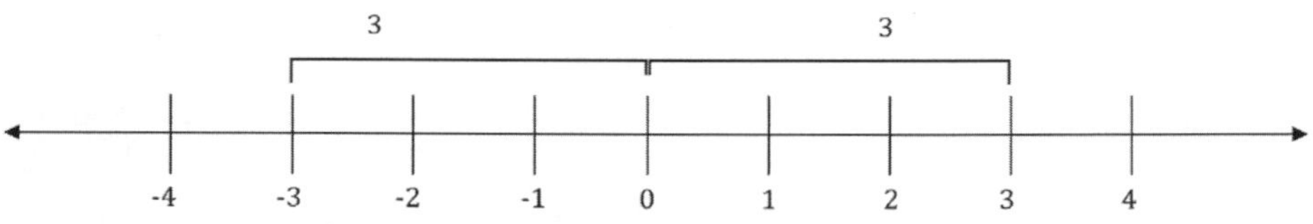

Review Video: Absolute Value
Visit mometrix.com/academy and enter code: 314669

Operations

An **operation** is simply a mathematical process that takes some value(s) as input(s) and produces an output. Elementary operations are often written in the following form: *value operation value*. For instance, in the expression $1 + 2$ the values are 1 and 2 and the operation is addition. Performing the operation gives the output of 3. In this way we can say that $1 + 2$ and 3 are equal, or $1 + 2 = 3$.

ADDITION

Addition increases the value of one quantity by the value of another quantity (both called **addends**). Example: $2 + 4 = 6$ or $8 + 9 = 17$. The result is called the **sum**. With addition, the order does not matter, $4 + 2 = 2 + 4$.

When adding signed numbers, if the signs are the same simply add the absolute values of the addends and apply the original sign to the sum. For example, $(+4) + (+8) = +12$ and $(-4) + (-8) = -12$. When the original signs are different, take the absolute values of the addends and subtract the smaller value from the larger value, then apply the original sign of the larger value to the difference. Example: $(+4) + (-8) = -4$ and $(-4) + (+8) = +4$.

SUBTRACTION

Subtraction is the opposite operation to addition; it decreases the value of one quantity (the **minuend**) by the value of another quantity (the **subtrahend**). For example, $6 - 4 = 2$ or $17 - 8 = 9$. The result is called the **difference**. Note that with subtraction, the order does matter, $6 - 4 \neq 4 - 6$.

For subtracting signed numbers, change the sign of the subtrahend and then follow the same rules used for addition. Example: $(+4) - (+8) = (+4) + (-8) = -4$

MULTIPLICATION

Multiplication can be thought of as repeated addition. One number (the **multiplier**) indicates how many times to add the other number (the **multiplicand**) to itself. Example: $3 \times 2 = 2 + 2 + 2 = 6$. With multiplication, the order does not matter, $2 \times 3 = 3 \times 2$ or $3 + 3 = 2 + 2 + 2$, either way the result (the **product**) is the same.

If the signs are the same, the product is positive when multiplying signed numbers. Example: $(+4) \times (+8) = +32$ and $(-4) \times (-8) = +32$. If the signs are opposite, the product is negative. Example: $(+4) \times (-8) = -32$ and $(-4) \times (+8) = -32$. When more than two factors are multiplied together, the sign of the product is determined by how many negative factors are present. If there are an odd number of negative factors then the product is negative, whereas an even number of negative factors indicates a positive product. Example: $(+4) \times (-8) \times (-2) = +64$ and $(-4) \times (-8) \times (-2) = -64$.

DIVISION

Division is the opposite operation to multiplication; one number (the **divisor**) tells us how many parts to divide the other number (the **dividend**) into. The result of division is called the **quotient**. Example: $20 \div 4 = 5$. If 20 is split into 4 equal parts, each part is 5. With division, the order of the numbers does matter, $20 \div 4 \neq 4 \div 20$.

The rules for dividing signed numbers are similar to multiplying signed numbers. If the dividend and divisor have the same sign, the quotient is positive. If the dividend and divisor have opposite signs, the quotient is negative. Example: $(-4) \div (+8) = -0.5$.

Review Video: Mathematical Operations
Visit mometrix.com/academy and enter code: 208095

PARENTHESES

Parentheses are used to designate which operations should be done first when there are multiple operations. Example: $4 - (2 + 1) = 1$; the parentheses tell us that we must add 2 and 1, and then subtract the sum from 4, rather than subtracting 2 from 4 and then adding 1 (this would give us an answer of 3).

Review Video: Mathematical Parentheses
Visit mometrix.com/academy and enter code: 978600

EXPONENTS

An **exponent** is a superscript number placed next to another number at the top right. It indicates how many times the base number is to be multiplied by itself. Exponents provide a shorthand way to write what would be a longer mathematical expression. Example: $2^4 = 2 \times 2 \times 2 \times 2$. A number with an exponent of 2 is said to be "squared," while a number with an exponent of 3 is said to be "cubed." The value of a number raised to an exponent is called its power. So 8^4 is read as "8 to the 4th power," or "8 raised to the power of 4."

> **Review Video: Exponents**
> Visit mometrix.com/academy and enter code: 600998

Roots

A **root**, such as a square root, is another way of writing a fractional exponent. Instead of using a superscript, roots use the radical symbol ($\sqrt{\ }$) to indicate the operation. A radical will have a number underneath the bar, and may sometimes have a number in the upper left: $\sqrt[n]{a}$, read as "the nth root of a." The relationship between radical notation and exponent notation can be described by this equation:

$$\sqrt[n]{a} = a^{\frac{1}{n}}$$

The two special cases of $n = 2$ and $n = 3$ are called square roots and cube roots. If there is no number to the upper left, the radical is understood to be a square root ($n = 2$). Nearly all of the roots you encounter will be square roots. A square root is the same as a number raised to the one-half power. When we say that a is the square root of b ($a = \sqrt{b}$), we mean that a multiplied by itself equals b: ($a \times a = b$).

A **perfect square** is a number that has an integer for its square root. There are 10 perfect squares from 1 to 100: 1, 4, 9, 16, 25, 36, 49, 64, 81, 100 (the squares of integers 1 through 10).

> **Review Video: Roots**
> Visit mometrix.com/academy and enter code: 795655
>
> **Review Video: Perfect Squares and Square Roots**
> Visit mometrix.com/academy and enter code: 648063

Word Problems and Mathematical Symbols

When working on word problems, you must be able to translate verbal expressions or "math words" into math symbols. This chart contains several "math words" and their appropriate symbols:

Phrase	Symbol
equal, is, was, will be, has, costs, gets to, is the same as, becomes	$=$
times, of, multiplied by, product of, twice, doubles, halves, triples	$\times$
divided by, per, ratio of/to, out of	$\div$
plus, added to, sum, combined, and, more than, totals of	$+$
subtracted from, less than, decreased by, minus, difference between	$-$
what, how much, original value, how many, a number, a variable	x, n, etc.

> **Review Video: Understanding Word Problems**
> Visit mometrix.com/academy and enter code: 499199

Examples of Translated Mathematical Phrases

- The phrase four more than twice a number can be written algebraically as $2x + 4$.
- The phrase half a number decreased by six can be written algebraically as $\frac{1}{2}x - 6$.
- The phrase the sum of a number and the product of five and that number can be written algebraically as $x + 5x$.

- You may see a test question that says, "Olivia is constructing a bookcase from seven boards. Two of them are for vertical supports and five are for shelves. The height of the bookcase is twice the width of the bookcase. If the seven boards total 36 feet in length, what will be the height of Olivia's bookcase?" You would need to make a sketch and then create the equation to determine the width of the shelves. The height can be represented as double the width. (If x represents the width of the shelves in feet, then the height of the bookcase is $2x$. Since the seven boards total 36 feet, $2x + 2x + x + x + x + x + x = 36$ or $9x = 36$; $x = 4$. The height is twice the width, or 8 feet.)

Subtraction with Regrouping

A great way to make use of some of the features built into the decimal system would be regrouping when attempting longform subtraction operations. When subtracting within a place value, sometimes the minuend is smaller than the subtrahend; **regrouping** enables you to 'borrow' a unit from a place value to the left in order to get a positive difference. For example, consider subtracting 189 from 525 with regrouping.

First, set up the subtraction problem in vertical form:

$$\begin{array}{r} 525 \\ -\ 189 \\ \hline \end{array}$$

Notice that the numbers in the ones and tens columns of 525 are smaller than the numbers in the ones and tens columns of 189. This means you will need to use regrouping to perform subtraction:

$$\begin{array}{rrrr} & 5 & 2 & 5 \\ - & 1 & 8 & 9 \\ \hline \end{array}$$

To subtract 9 from 5 in the ones column you will need to borrow from the 2 in the tens column:

$$\begin{array}{rrrr} & 5 & 1 & 15 \\ - & 1 & 8 & 9 \\ \hline & & & 6 \end{array}$$

Next, to subtract 8 from 1 in the tens column you will need to borrow from the 5 in the hundreds column:

$$\begin{array}{rrrr} & 4 & 11 & 15 \\ - & 1 & 8 & 9 \\ \hline & & 3 & 6 \end{array}$$

Last, subtract the 1 from the 4 in the hundreds column:

$$\begin{array}{rrrr} & 4 & 11 & 15 \\ - & 1 & 8 & 9 \\ \hline & 3 & 3 & 6 \end{array}$$

Review Video: Subtracting Large Numbers
Visit mometrix.com/academy and enter code: 603350

Order of Operations

The **order of operations** is a set of rules that dictates the order in which we must perform each operation in an expression so that we will evaluate it accurately. If we have an expression that includes multiple different operations, the order of operations tells us which operations to do first. The most common mnemonic for the order of operations is **PEMDAS**, or "Please Excuse My Dear Aunt Sally." PEMDAS stands for parentheses, exponents, multiplication, division, addition, and subtraction. It is important to understand that multiplication and division have equal precedence, as do addition and subtraction, so those pairs of operations are simply worked from left to right in order.

For example, evaluating the expression $5 + 20 \div 4 \times (2 + 3)^2 - 6$ using the correct order of operations would be done like this:

- **P:** Perform the operations inside the parentheses: $(2 + 3) = 5$
- **E:** Simplify the exponents: $(5)^2 = 5 \times 5 = 25$
  - The expression now looks like this: $5 + 20 \div 4 \times 25 - 6$
- **MD:** Perform multiplication and division from left to right: $20 \div 4 = 5$; then $5 \times 25 = 125$
  - The expression now looks like this: $5 + 125 - 6$
- **AS:** Perform addition and subtraction from left to right: $5 + 125 = 130$; then $130 - 6 = 124$

Review Video: Order of Operations
Visit mometrix.com/academy and enter code: 259675

Properties of Operations

The Commutative Property

The commutative property applies to addition and multiplication and states that these operations can be completed in any order. The **commutative property of addition** states that numbers and terms can be added together in any order to still get the same value. For example, $3 + 4 = 7$ and $4 + 3 = 7$. Also, we can use the commutative property of addition to show that $3x + 4 + 2^2$ is equivalent to $4 + 3x + 2^2$ and $2^2 + 4 + 3x$. When adding terms, you can add in any order and get the same value.

The **commutative property of multiplication** states that numbers and terms can be multiplied in any order to get the same value. For example, 12×3 is equivalent to 3×12. Additionally, we can use the commutative property of multiplication to assume that $(5 + 3) \times (36 - 6)$ is equivalent to $(36 - 6) \times (5 + 3)$. You can multiply terms in any order and still get the same value.

The Associative Property

The **associative property of addition** states that if three or more terms are being added together, the value is the same regardless of the groupings.

For example, given the expression $3 + 4 + 6$, these terms can be grouped and added in any form. $3 + 4 + 6$ is equivalent to $(3 + 4) + 6$ and is also equivalent to $3 + (4 + 6)$. This can be applied to write equivalent expressions in a variety of ways.

For example, suppose we are given the expression $5 + (y + 2) + 4$. We can generate equivalent expressions knowing the associative property. Knowing that when three or more terms are added, the grouping is irrelevant, we can say that this expression is equivalent to $5 + y + (2 + 4)$, and it is equivalent to $(5 + y) + (2 + 4)$. It is even equivalent to $5 + y + 2 + 4$.

The **associative property of multiplication** states that if three or more terms are being multiplied together, the value is the same regardless of the grouping. We can use this property to identify and generate equivalent expressions.

For example, given the expression $2 \times 7 \times 3$, these terms can be grouped in any way and still get the same value. $2 \times 7 \times 3$ is equivalent to $(2 \times 7) \times 3$ or $2 \times (7 \times 3)$.

The Identity Property

The **identity property of multiplication** states that when a number is multiplied by 1, you get the same number. That is, anything multiplied by 1 is itself. For example, $2 \times 1 = 2$, or $1 \times -36 = -36$. Using the identity property of multiplication, we can identify and generate equivalent expressions. Let's say that we are given the expression $15 - (3 \times 4)$. We can generate equivalent expressions using the identity property. One equivalent expression example would be $(15 \times 1) - (3 \times 4)$. Another example would be $15 - (1 \times 3 \times 4)$. We can say these expressions are equivalent because the identity property of multiplication states that we can multiply any portion of an expression by 1 to get the same value.

The **identity property of addition** states that when 0 is added to a number, you get the same number. For example, $2 + 0 = 2$, or $0 + -3 = -3$. We can also use this property to identify and generate equivalent expressions. For example, if we are given the expression $2 \times (1 + 2)$, we could write the equivalent expressions $2 \times (0 + 1 + 2)$ or $(2 + 0) \times (1 + 2)$.

The Inverse Property

The **inverse property of addition** states that the sum of a number and its opposite is always equal to 0. Remember, the opposite of a number is a number that is opposite on the number line from zero, or the same number with the opposite sign. For example, −4 is opposite to 4, and 1,726.9 is opposite to −1,726.9. So, the inverse property of addition states that if you add opposite numbers, their sum is zero. For example, $5 + (-5) = 0$ and $-5 + 5 = 0$.

The **inverse property of multiplication** states that a number multiplied by its reciprocal is always equal to 1. The **reciprocal** of a number is its "flipped" fraction. For example, the reciprocal of 5 is $\frac{1}{5}$, or the reciprocal of $\frac{2}{3}$ is $\frac{3}{2}$. The inverse property of multiplication can be applied for these values, $5 \times \frac{1}{5} = 1$ and $\frac{2}{3} \times \frac{3}{2} = 1$. This is because when you multiply across, you get a fraction that is equal to 1.

$$\frac{2}{3} \times \frac{3}{2} = \frac{6}{6} = 1$$

The Distributive Property

The **distributive property** explains how multiplication and addition interact. It says that when multiplying one number by the sum of two other numbers, the same result can also be obtained by multiplying the one number by each of the numbers individually and then adding the products. For example, to multiply 2 by the sum of 7 and 3, the direct approach says, "the sum of 7 and 3 is 10, and 2 times 10 is 20." This would be expressed as $2 \times (7 + 3) = 2 \times 10 = 20$. On the other hand, the distributive property states that the same answer can be achieved by multiplying each number inside the parentheses and adding the products. That is, "the product of 2 and 7 is 14, the product of

2 and 3 is 6, and the sum of 14 and 6 is 20." This would be expressed as $2 \times (7 + 3) = 2 \times 7 + 2 \times 3 = 14 + 6 = 20$, and it is demonstrated below.

$$2 \times (7 + 3) = 2 \times 7 + 2 \times 3$$

This same concept can be used when multiplying a number by the difference of two numbers. For example, $5 \times (10 - 4) = 5 \times 10 - 5 \times 4$. Since $5 \times 10 = 50$ and $5 \times 4 = 20$, the result is $50 - 20 = 30$. This answer can be checked by subtracting inside the parentheses first and then multiplying: $5 \times (10 - 4) = 5 \times 6 = 30$.

Review Video: Commutative, Associative, and Distributive Properties
Visit mometrix.com/academy and enter code: 483176

Properties of Exponents

The properties of exponents are as follows:

Property	Description
$a^1 = a$	Any number to the power of 1 is equal to itself
$1^n = 1$	The number 1 raised to any power is equal to 1
$a^0 = 1$	Any number raised to the power of 0 is equal to 1
$a^n \times a^m = a^{n+m}$	Add exponents to multiply powers of the same base number
$a^n \div a^m = a^{n-m}$	Subtract exponents to divide powers of the same base number
$(a^n)^m = a^{n \times m}$	When a power is raised to a power, the exponents are multiplied
$(a \times b)^n = a^n \times b^n$ $(a \div b)^n = a^n \div b^n$	Multiplication and division operations inside parentheses can be raised to a power. This is the same as each term being raised to that power.
$a^{-n} = \frac{1}{a^n}$	A negative exponent is the same as the reciprocal of a positive exponent

Note that exponents do not have to be integers. Fractional or decimal exponents follow all the rules above as well. Example: $5^{\frac{1}{4}} \times 5^{\frac{3}{4}} = 5^{\frac{1}{4}+\frac{3}{4}} = 5^1 = 5$.

Review Video: Properties of Exponents
Visit mometrix.com/academy and enter code: 532558

Factors and Multiples

Factors and Greatest Common Factor

A whole number a is a **factor** (or **divisor**) of a whole number b if a divides b evenly. In other words, a is a factor of b if the quotient $b \div a$ is a whole number with a remainder of 0. For instance, 3 is a factor of 12 because $12 \div 3 = 4$ with no remainder. Another way to say this is that a is a factor of b if we can multiply a by another whole number to get b. So, we can also show that 3 is a factor of 12 by noting that $3 \times 4 = 12$.

Every positive whole number has 1 and itself as factors. If a whole number greater than one has *only* 1 and itself as factors, we call it a **prime number**. For instance, 5 is a prime number because its only factors are 1 and 5. The first several prime numbers are 2, 3, 5, 7, 11, and 13.

If a whole number greater than 1 is not prime—that is, if it has factors besides 1 and itself—then it is a **composite number.** For instance, 10 is a composite number because it has factors 2 and 5 in addition to 1 and 10. The first several composite numbers are 4, 6, 8, 9, 10, 12, 14, and 15.

A **prime factor** of a whole number is a factor that is also a prime number. For example, the prime factors of 12 are 2 and 3. The prime factors of 15 are 3 and 5.

A **common factor** of two (or more) whole numbers is a number that is a factor of both (or all) of them. For example, the factors of 12 are 1, 2, 3, 4, 6, and 12, while the factors of 15 are 1, 3, 5, and 15. The common factors (underlined) of 12 and 15 are 1 and 3.

The **greatest common factor** (**GCF**) of two (or more) whole numbers is the largest number that is a factor of both (or all) of them. For example, the factors of 15 are 1, 3, 5, and 15; the factors of 35 are 1, 5, 7, and 35. Therefore, the greatest common factor of 15 and 35 is 5.

Review Video: Factors
Visit mometrix.com/academy and enter code: 920086

Review Video: Prime Numbers and Factorization
Visit mometrix.com/academy and enter code: 760669

Multiples and Least Common Multiple

A whole number b is a **multiple** of a whole number a when a is a factor of b. This means that b is the product of a and another whole number. For example, the multiples of 7 are $0 \times 7 = 0, 1 \times 7 = 7, 2 \times 7 = 14, 3 \times 7 = 21, 4 \times 7 = 28, 5 \times 7 = 35, \ldots$. Dividing 0, 7, 14, 21, 28, and 35 by 7 results in the whole numbers 0, 1, 2, 3, 4, and 5, respectively, showing that 7 is a factor of these numbers.

The least common multiple (**LCM**) of two (or more) whole numbers is the smallest number that is a multiple of both (or all) of them. For example, the multiples of 3 are 3, 6, 9, 12, 15, ...; the multiples of 5 are 5, 10, 15, 20, ... The smallest number that appears in both lists is 15, so the least common multiple of 3 and 5 is 15.

Review Video: Multiples
Visit mometrix.com/academy and enter code: 626738

Review Video: Greatest Common Factor and Least Common Multiple
Visit mometrix.com/academy and enter code: 838699

Fractions

A **fraction** is a number that is expressed as one integer written above another integer, with a dividing line between them $\left(\frac{x}{y}\right)$. It represents the **quotient** of the two numbers "x divided by y." It can also be thought of as x out of y equal parts.

The top number of a fraction is called the **numerator**, and it represents the number of parts under consideration. The 1 in $\frac{1}{4}$ means that 1 part out of the whole is being considered in the calculation.

The bottom number of a fraction is called the **denominator**, and it represents the total number of equal parts. The 4 in $\frac{1}{4}$ means that the whole consists of 4 equal parts. A fraction cannot have a denominator of zero; this is referred to as "*undefined*."

Fractions can be manipulated, without changing the value of the fraction, by multiplying or dividing (but not adding or subtracting) both the numerator and denominator by the same number. If you divide both numbers by a common factor, you are **reducing** or simplifying the fraction. Two fractions that have the same value but are expressed differently are known as **equivalent fractions**. For example, $\frac{2}{10}, \frac{3}{15}, \frac{4}{20}$, and $\frac{5}{25}$ are all equivalent fractions. They can also all be reduced or simplified to $\frac{1}{5}$.

When two fractions are manipulated so that they have the same denominator, this is known as finding a **common denominator**. The number chosen to be that common denominator should be the least common multiple of the two original denominators. Example: $\frac{3}{4}$ and $\frac{5}{6}$; the least common multiple of 4 and 6 is 12. Manipulating to achieve the common denominator: $\frac{3}{4} = \frac{9}{12}; \frac{5}{6} = \frac{10}{12}$.

Review Video: Overview of Fractions
Visit mometrix.com/academy and enter code: 262335

Proper Fractions and Mixed Numbers

A fraction whose denominator is greater than its numerator is known as a **proper fraction**, while a fraction whose numerator is greater than its denominator is known as an **improper fraction**. Proper fractions have values *less than one* and improper fractions have values *greater than one.*

A **mixed number** is a number that contains both an integer and a fraction. Any improper fraction can be rewritten as a mixed number. Example: $\frac{8}{3} = \frac{6}{3} + \frac{2}{3} = 2 + \frac{2}{3} = 2\frac{2}{3}$. Similarly, any mixed number can be rewritten as an improper fraction. Example: $1\frac{3}{5} = 1 + \frac{3}{5} = \frac{5}{5} + \frac{3}{5} = \frac{8}{5}$.

Review Video: Proper and Improper Fractions and Mixed Numbers
Visit mometrix.com/academy and enter code: 211077

Adding and Subtracting Fractions

If two fractions have a common denominator, they can be added or subtracted simply by adding or subtracting the two numerators and retaining the same denominator. If the two fractions do not already have the same denominator, one or both of them must be manipulated to achieve a common denominator before they can be added or subtracted. Example: $\frac{1}{2} + \frac{1}{4} = \frac{2}{4} + \frac{1}{4} = \frac{3}{4}$.

Review Video: Adding and Subtracting Fractions
Visit mometrix.com/academy and enter code: 378080

Multiplying Fractions

Two fractions can be multiplied by multiplying the two numerators to find the new numerator and the two denominators to find the new denominator. Example: $\frac{1}{3} \times \frac{2}{3} = \frac{1\times2}{3\times3} = \frac{2}{9}$.

Dividing Fractions

Two fractions can be divided by flipping the numerator and denominator of the second fraction and then proceeding as though it were a multiplication problem. Example: $\frac{2}{3} \div \frac{3}{4} = \frac{2}{3} \times \frac{4}{3} = \frac{8}{9}$.

Review Video: Multiplying and Dividing Fractions
Visit mometrix.com/academy and enter code: 473632

Multiplying a Mixed Number by a Whole Number or a Decimal

When multiplying a mixed number by something, it is usually best to convert it to an improper fraction first. Additionally, if the multiplicand is a decimal, it is most often simplest to convert it to a fraction. For instance, to multiply $4\frac{3}{8}$ by 3.5, begin by rewriting each quantity as a whole number plus a proper fraction. Remember, a mixed number is a fraction added to a whole number and a decimal is a representation of the sum of fractions, specifically tenths, hundredths, thousandths, and so on:

$$4\frac{3}{8} \times 3.5 = \left(4 + \frac{3}{8}\right) \times \left(3 + \frac{1}{2}\right)$$

Next, the quantities being added need to be expressed with the same denominator. This is achieved by multiplying and dividing the whole number by the denominator of the fraction. Recall that a whole number is equivalent to that number divided by 1:

$$= \left(\frac{4}{1} \times \frac{8}{8} + \frac{3}{8}\right) \times \left(\frac{3}{1} \times \frac{2}{2} + \frac{1}{2}\right)$$

When multiplying fractions, remember to multiply the numerators and denominators separately:

$$= \left(\frac{4 \times 8}{1 \times 8} + \frac{3}{8}\right) \times \left(\frac{3 \times 2}{1 \times 2} + \frac{1}{2}\right)$$
$$= \left(\frac{32}{8} + \frac{3}{8}\right) \times \left(\frac{6}{2} + \frac{1}{2}\right)$$

Now that the fractions have the same denominators, they can be added:

$$= \frac{35}{8} \times \frac{7}{2}$$

Finally, perform the last multiplication and then simplify:

$$= \frac{35 \times 7}{8 \times 2} = \frac{245}{16} = \frac{240}{16} + \frac{5}{16} = 15\frac{5}{16}$$

Comparing Fractions

It is important to master the ability to compare and order fractions. This skill is relevant to many real-world scenarios. For example, carpenters often compare fractional construction nail lengths when preparing for a project, and bakers often compare fractional measurements to have the correct ratio of ingredients. There are three commonly used strategies when comparing fractions. These strategies are referred to as the common denominator approach, the decimal approach, and the cross-multiplication approach.

USING A COMMON DENOMINATOR TO COMPARE FRACTIONS

The fractions $\frac{2}{3}$ and $\frac{4}{7}$ have different denominators. $\frac{2}{3}$ has a denominator of 3, and $\frac{4}{7}$ has a denominator of 7. In order to precisely compare these two fractions, it is necessary to use a common denominator. A common denominator is a common multiple that is shared by both denominators. In this case, the denominators 3 and 7 share a multiple of 21. In general, it is most efficient to select the least common multiple for the two denominators.

Rewrite each fraction with the common denominator of 21. Then, calculate the new numerators as illustrated below.

$$\frac{2}{3} \overset{\times 7}{=} \frac{\mathbf{14}}{21} \qquad \frac{4}{7} \overset{\times 3}{=} \frac{\mathbf{12}}{21}$$

For $\frac{2}{3}$, multiply the numerator and denominator by 7. The result is $\frac{14}{21}$.

For $\frac{4}{7}$, multiply the numerator and denominator by 3. The result is $\frac{12}{21}$.

Now that both fractions have a denominator of 21, the fractions can accurately be compared by comparing the numerators. Since 14 is greater than 12, the fraction $\frac{14}{21}$ is greater than $\frac{12}{21}$. This means that $\frac{2}{3}$ is greater than $\frac{4}{7}$.

USING DECIMALS TO COMPARE FRACTIONS

Sometimes decimal values are easier to compare than fraction values. For example, $\frac{5}{8}$ is equivalent to 0.625 and $\frac{3}{5}$ is equivalent to 0.6. This means that the comparison of $\frac{5}{8}$ and $\frac{3}{5}$ can be determined by comparing the decimals 0.625 and 0.6. When both decimal values are extended to the thousandths place, they become 0.625 and 0.600, respectively. It becomes clear that 0.625 is greater than 0.600 because 625 thousandths is greater than 600 thousandths. In other words, $\frac{5}{8}$ is greater than $\frac{3}{5}$ because 0.625 is greater than 0.6.

USING CROSS-MULTIPLICATION TO COMPARE FRACTIONS

Cross-multiplication is an efficient strategy for comparing fractions. This is a shortcut for the common denominator strategy. Start by writing each fraction next to one another. Multiply the numerator of the fraction on the left by the denominator of the fraction on the right. Write down the result next to the fraction on the left. Now multiply the numerator of the fraction on the right by the denominator of the fraction on the left. Write down the result next to the fraction on the right. Compare both products. The fraction with the larger result is the larger fraction.

Consider the fractions $\frac{4}{7}$ and $\frac{5}{9}$.

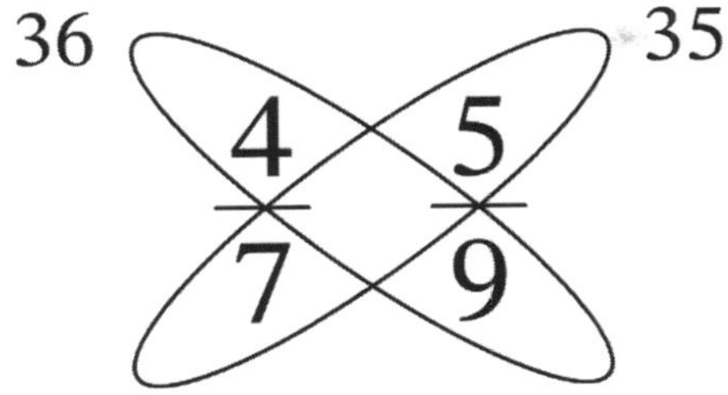

36 is greater than 35. Therefore, $\frac{4}{7}$ is greater than $\frac{5}{9}$.

Decimals

Decimals are one way to represent parts of a whole. Using the place value system, each digit to the right of a decimal point denotes the number of units of a corresponding *negative* power of ten. For example, consider the decimal 0.24. We can use a model to represent the decimal. Since a dime is worth one-tenth of a dollar and a penny is worth one-hundredth of a dollar, one possible model to represent this fraction is to have 2 dimes representing the 2 in the tenths place and 4 pennies representing the 4 in the hundredths place:

To write the decimal as a fraction, put the decimal in the numerator with 1 in the denominator. Multiply the numerator and denominator by tens until there are no more decimal places. Then simplify the fraction to lowest terms. For example, converting 0.24 to a fraction:

$$0.24 = \frac{0.24}{1} = \frac{0.24 \times 100}{1 \times 100} = \frac{24}{100} = \frac{6}{25}$$

Review Video: Decimals
Visit mometrix.com/academy and enter code: 837268

Operations with Decimals

Adding and Subtracting Decimals

When adding and subtracting decimals, the decimal points must always be aligned. Adding decimals is just like adding regular whole numbers. Example: $4.5 + 2.0 = 6.5$.

If the problem-solver does not properly align the decimal points, an incorrect answer of 4.7 may result. An easy way to add decimals is to align all of the decimal points in a vertical column visually. This will allow you to see exactly where the decimal should be placed in the final answer. Begin adding from right to left. Add each column in turn, making sure to carry the number to the left if a column adds up to more than 9. The same rules apply to the subtraction of decimals.

Review Video: Adding and Subtracting Decimals
Visit mometrix.com/academy and enter code: 381101

Multiplying Decimals

A simple multiplication problem has two components: a **multiplicand** and a **multiplier**. When multiplying decimals, work as though the numbers were whole rather than decimals. Once the final product is calculated, count the number of places to the right of the decimal in both the multiplicand and the multiplier. Then, count that number of places from the right of the product and place the decimal in that position.

For example, 12.3×2.56 has a total of three places to the right of the respective decimals. Multiply 123×256 to get 31,488. Now, beginning on the right, count three places to the left and insert the decimal. The final product will be 31.488.

Review Video: How to Multiply Decimals
Visit mometrix.com/academy and enter code: 731574

Dividing Decimals

Every division problem has a **divisor** and a **dividend**. The dividend is the number that is being divided. In the problem $14 \div 7$, 14 is the dividend and 7 is the divisor. In a division problem with decimals, the divisor must be converted into a whole number. Begin by moving the decimal in the divisor to the right until a whole number is created. Next, move the decimal in the dividend the same number of spaces to the right. For example, 4.9 into 24.5 would become 49 into 245. The decimal was moved one space to the right to create a whole number in the divisor, and then the same was done for the dividend. Once the whole numbers are created, the problem is carried out normally: $245 \div 49 = 5$.

Review Video: Dividing Decimals
Visit mometrix.com/academy and enter code: 560690

Review Video: Dividing Decimals by Whole Numbers
Visit mometrix.com/academy and enter code: 535669

Percentages

Percentages can be thought of as fractions that are based on a whole of 100; that is, one whole is equal to 100%. The word **percent** means "per hundred." Percentage problems are often presented in three main ways:

- Find what percentage of some number another number is.
  - Example: What percentage of 40 is 8?
- Find what number is some percentage of a given number.
  - Example: What number is 20% of 40?
- Find what number another number is a given percentage of.
  - Example: What number is 8 20% of?

There are three components in each of these cases: a **whole** (W), a **part** (P), and a **percentage** (%). These are related by the equation: $P = W \times \%$. This can easily be rearranged into other forms that may suit different questions better: $\% = \frac{P}{W}$ and $W = \frac{P}{\%}$. Percentage problems are often also word problems. As such, a large part of solving them is figuring out which quantities are what. For example, consider the following word problem:

In a school cafeteria, 7 students choose pizza, 9 choose hamburgers, and 4 choose tacos. What percentage of students choose tacos?

To find the whole, you must first add all of the parts: $7 + 9 + 4 = 20$. The percentage can then be found by dividing the part by the whole $\left(\% = \frac{P}{W}\right)$: $\frac{4}{20} = \frac{20}{100} = 20\%$.

Review Video: Computation with Percentages
Visit mometrix.com/academy and enter code: 693099

Calculating Percent Change

Suppose a quantity has a particular value (the *old value*) and then we add something (the *change*) to it to get another value (the *new value*). We can describe this process by the simple equation $(\text{old value}) + \text{change} = (\text{new value})$. If we know the old and new values, we can rearrange this equation to find the change, getting $\text{change} = (\text{new value}) - (\text{old value})$. For instance, if a store's price for a box of computer paper goes from \$20 last week to \$25 this week, this is a change of $(\text{new value}) - (\text{old value}) = \$25 - \$20 = \$5$. Or, if the size of the freshman class at a college goes from 500 students one year to 440 students the next year, this is a change of $(\text{new value}) - (\text{old value}) = 440 - 500 = -60$ students. So, we see that change can be positive or negative.

Instead of the word *change*, we sometimes use the words *increase* or *decrease* to specify whether the value goes up or down, respectively. In the examples above, the price of computer paper increases by \$5 and the freshman class decreases by 60 students. Note that the decrease is 60 students and not -60 because the word *decrease* already means that the value goes down. So, *increase* is the same as positive change and *decrease* is the opposite or negative change.

If the changing quantity represents an amount (how much of something there is), we can also calculate the **percent change**. This is the change expressed as a percentage of the old amount. To calculate this, we divide the change by the old amount and express the quotient as a percent. That is, we use the formula $\text{percent change} = \frac{\text{change}}{\text{old value}}$, converting the resulting decimal answer to a percent. In the examples above, the price of a box of computer paper has a percent change of $\frac{\text{change in price}}{\text{old price}} = \frac{\$5}{\$20} = 0.25 = 25\%$, and the size of the freshman class at the college has a percent change of $\frac{\text{change in enrollment}}{\text{old enrollment}} = \frac{-60}{500} = -0.12 = -12\%$. We can also use the terms *percent increase* and *percent decrease*, saying that the price of computer paper increases by 25% and the size of the freshman class decreases by 12%. Note that the denominator is always the old amount, never the new amount.

Example: Your landlord raises your rent from \$1,500 to \$1,700 per month. To find the percent change in your rent (rounded to the nearest tenth of a percent), you calculate as follows.

$$\text{percent change in rent} = \frac{\text{change in rent}}{\text{old rent}} = \frac{(\text{new rent}) - (\text{old rent})}{\text{old rent}}$$
$$= \frac{\$1{,}700 - \$1{,}500}{\$1{,}500} = \frac{\$200}{\$1{,}500} = 0.1333\ldots \approx 13.3\%$$

Therefore, the percent change in your rent is approximately 13.3%.

Review Video: Percent Change
Visit mometrix.com/academy and enter code: 907890

Converting Between Percentages, Fractions, and Decimals

Converting decimals to percentages and percentages to decimals is as simple as moving the decimal point. To *convert from a decimal to a percentage*, move the decimal point **two places to the right**. To *convert from a percentage to a decimal*, move it **two places to the left**. It may be helpful to remember that the percentage number will always be larger than the equivalent decimal number. Example:

$$0.23 = 23\% \qquad 5.34 = 534\% \qquad 0.007 = 0.7\%$$
$$700\% = 7.00 \qquad 86\% = 0.86 \qquad 0.15\% = 0.0015$$

To convert a fraction to a decimal, simply divide the numerator by the denominator in the fraction. To convert a decimal to a fraction, put the decimal in the numerator with 1 in the denominator. Multiply the numerator and denominator by tens until there are no more decimal places. Then simplify the fraction to lowest terms. For example, converting 0.24 to a fraction:

$$0.24 = \frac{0.24}{1} = \frac{0.24 \times 100}{1 \times 100} = \frac{24}{100} = \frac{6}{25}$$

Fractions can be converted to a percentage by finding equivalent fractions with a denominator of 100. Example:

$$\frac{7}{10} = \frac{70}{100} = 70\% \quad \frac{1}{4} = \frac{25}{100} = 25\%$$

To convert a percentage to a fraction, divide the percentage number by 100 and reduce the fraction to its simplest possible terms. Example:

$$60\% = \frac{60}{100} = \frac{3}{5} \quad 96\% = \frac{96}{100} = \frac{24}{25}$$

Review Video: Converting Fractions to Percentages and Decimals
Visit mometrix.com/academy and enter code: 306233

Review Video: Converting Percentages to Decimals and Fractions
Visit mometrix.com/academy and enter code: 287297

Review Video: Converting Decimals to Fractions and Percentages
Visit mometrix.com/academy and enter code: 986765

Review Video: Converting Decimals, Improper Fractions, and Mixed Numbers
Visit mometrix.com/academy and enter code: 696924

Rational and Irrational Numbers

The term **rational** means that the number can be expressed as a ratio or fraction. That is, a number, r, is rational if and only if it can be represented by a fraction $\frac{a}{b}$ where a and b are integers and b does not equal 0. The set of rational numbers includes integers and decimals. If there is no finite way to represent a value with a fraction of integers, then the number is **irrational**. Common irrational numbers are π and the square roots of whole numbers that are not perfect squares (e.g., $\sqrt{5}$ or $\sqrt{21}$). The sum or product of an integer and an irrational number is always irrational (e.g., 3π or $7 + \sqrt{6}$).

Review Video: Rational and Irrational Numbers
Visit mometrix.com/academy and enter code: 280645

Review Video: Ordering Rational Numbers
Visit mometrix.com/academy and enter code: 419578

Review Video: Irrational Numbers on a Number Line
Visit mometrix.com/academy and enter code: 433866

Chapter Quiz

Ready to see how well you retained what you just read? Scan the QR code to go directly to the chapter quiz interface for this study guide. If you're using a computer, simply visit the online resources page at **mometrix.com/resources719/orelamgmath** and click the Chapter Quizzes link.

Algebra and Functions

Proportionality

Proportions and Ratios

Proportions

There is a **proportion** between two variable quantities if there is a constant relationship between their products or quotients, a relationship that does not change as the quantities themselves change.

Given variable quantities x and y, we say that they are **directly proportional** (or that y **varies directly with** x) if their quotient or *ratio* is constant—that is, if there is a constant k such that $\frac{y}{x} = k$ is always true. Another way of saying this is that y is a constant multiple of x, so that $y = kx$ is always true. We call the number k the **constant of proportionality**. For example, if you drive at a constant 50 miles per hour, then the distance, y, that you travel in miles is 50 times the number of hours, x, that you drive. In symbols, $y = 50x$ miles (or $\frac{y}{x} = 50$ mph). So, the distance you travel, y, is directly proportional to (or varies directly with) the time you travel, x, with constant of proportionality $k = 50$ mph.

The quantities x and y are **inversely proportional** (or y varies inversely with x) if their product is constant—that is, if there is a constant k such that $xy = k$ is always true. Another way of saying this is to say that y is a constant multiple of the reciprocal of x so that $y = \frac{k}{x}$ is always true. For instance, suppose you drive at speed (rate) y mph for x hours, going a total of 120 miles. Since rate × time = distance, we get $xy = 120$ miles (or $y = \frac{120}{x}$ miles per hour). Thus, your driving speed, y, is inversely proportional to (or varies inversely with) your drive time, x, with constant of proportionality $k = 120$ miles.

Review Video: Proportions
Visit mometrix.com/academy and enter code: 505355

Ratios

A **ratio** expresses the sizes of two quantities relative to each other. For instance, suppose we have 3 copies of sheet music to share among 6 singers. We can divide the singers into groups of 2 and give each group 1 copy of the music. Thus, there is 1 copy of the music for every 2 singers, and we say that the **ratio** of sheet music to singers is 1 to 2, which we write either as a fraction $\frac{1}{2}$ or using a colon $1 : 2$. Of course, it is also true there are 3 copies for every 6 singers so that the ratio of sheet music to singers is also 3 to 6, which we write as $\frac{3}{6}$ or $3 : 6$. So, the ratios $\frac{1}{2}$ and $\frac{3}{6}$ express the same relative quantities of music and singers. We say that these ratios are equal or **equivalent**, and we note that ratios are equal precisely when their fractions are equal (so, in this case, $\frac{1}{2} = \frac{3}{6}$ as fractions). We can also express the quantities in the other order and say that the ratio of singers to music is $\frac{2}{1}$ or $2 : 1$ (or $\frac{6}{3}$ or $6 : 3$).

Review Video: Ratios
Visit mometrix.com/academy and enter code: 996914

Constant of Proportionality

If variable quantities x and y are proportional and we know a pair of corresponding values for them, then we can find their constant of proportionality. If they are directly proportional, we use the formula $\frac{y}{x} = k$. If they are inversely proportional, we use the formula $xy = k$.

Example: The cost in dollars, y, of buying fence posts is directly proportional to the number, x, that you buy. If it costs \$51 to buy 17 fence posts, what is the constant of proportionality? Because of direct proportionality, we know that $\frac{y}{x} = k$. Since this works for every pair of corresponding x- and y-values, it also works for $x = 17$ and $y = 51$. This gives us $\frac{51}{17} = k$, which simplifies to $k = 3$. Note also that this is the unit price, namely \$3 per fence post.

Work/Unit Rate

Unit rate expresses a quantity of one thing in terms of one unit of another. For example, if you travel 30 miles every two hours, a unit rate expresses this comparison in terms of one hour: in one hour you travel 15 miles, so your unit rate is 15 miles per hour. Other examples are how much one ounce of food costs (price per ounce) or figuring out how much one egg costs out of the dozen (price per 1 egg, instead of price per 12 eggs). The denominator of a unit rate is always 1. Unit rates are used to compare different situations to solve problems. For example, to make sure you get the best deal when deciding which kind of soda to buy, you can find the unit rate of each. If soda #1 costs \$1.50 for a 1-liter bottle, and soda #2 costs \$2.75 for a 2-liter bottle, it would be a better deal to buy soda #2, because its unit rate is only \$1.375 per liter, which is cheaper than soda #1. Unit rates can also help determine the length of time a given event will take. For example, if you can paint 2 rooms in 4.5 hours, you can determine how long it will take you to paint 5 rooms by solving for the unit rate per room and then multiplying that by 5.

Review Video: Rates and Unit Rates
Visit mometrix.com/academy and enter code: 185363

Cross Multiplication

Finding an Unknown in Equivalent Expressions

It is often necessary to apply information given about a rate or proportion to a new scenario. For example, if you know that Jedha can run a marathon (26.2 miles) in 3 hours, how long would it take her to run 10 miles at the same pace? Start by setting up equivalent expressions:

$$\frac{26.2 \text{ mi}}{3 \text{ hr}} = \frac{10 \text{ mi}}{x \text{ hr}}$$

Now, cross multiply and solve for x:

$$26.2x = 30$$
$$x = \frac{30}{26.2} = \frac{15}{13.1}$$
$$x \approx 1.15 \text{ hrs } or \text{ } 1 \text{ hr } 9 \text{ min}$$

So, at this pace, Jedha could run 10 miles in about 1.15 hours or about 1 hour and 9 minutes.

Review Video: Cross Multiplying Fractions
Visit mometrix.com/academy and enter code: 893904

Linear Expressions and Equations

Linear Expressions

Terms and Coefficients

Mathematical expressions consist of a combination of one or more values arranged in terms that are added together. As such, an expression could be just a single number, including zero. A **variable term** is the product of a real number, also called a **coefficient**, and one or more variables, each of which may be raised to an exponent. Expressions may also include numbers without a variable, called **constants** or **constant terms**. The expression $6s^2$, for example, is a single term where the coefficient is the real number 6 and the variable term is s^2. Note that if a term is written as simply a variable to some exponent, like t^2, then the coefficient is 1, because $t^2 = 1t^2$.

Linear Expressions

A **single variable linear expression** is the sum of a single variable term, where the variable has no exponent, and a constant, which may be zero. For instance, the expression $2w + 7$ has $2w$ as the variable term and 7 as the constant term. It is important to realize that terms are separated by addition or subtraction. Since an expression is a sum of terms, expressions such as $5x - 3$ can be written as $5x + (-3)$ to emphasize that the constant term is negative. A real-world example of a single variable linear expression is the perimeter of a square, four times the side length, often expressed: $4s$.

In general, a **linear expression** is the sum of any number of variable terms so long as none of the variables have an exponent and none of the terms have two variables multiplied together. For example, $3m + 8n - \frac{1}{4}p + 5.5q - 1$ is a linear expression, but $3y^3$ and $5xy$ are not. In the same way, the expression for the perimeter of a general triangle $(a + b + c)$ is linear, but the expression for the area of a square (s^2) is not.

Slope

Finding Slope Given Graph or Table

On a graph with two points, (x_1, y_1) and (x_2, y_2), the **slope** is found with the formula $m = \frac{y_2 - y_1}{x_2 - x_1}$; where $x_1 \neq x_2$ and m stands for slope. If the value of the slope is **positive**, the line has an *upward direction* from left to right. If the value of the slope is **negative**, the line has a *downward direction* from left to right. Consider the following example:

A new book goes on sale in bookstores and online stores. In the first month, 5,000 copies of the book are sold. Over time, the book continues to grow in popularity. The data for the number of copies sold is in the table below.

# of Months on Sale	1	2	3	4	5
# of Copies Sold (In Thousands)	5	10	15	20	25

So, the number of copies that are sold and the time that the book is on sale is a proportional relationship. In this example, an equation can be used to show the data: $y = 5x$, where x is the number of months that the book is on sale. Also, y is the number of copies sold. So, the slope of the corresponding line is $\frac{\text{rise}}{\text{run}} = \frac{5}{1} = 5$.

Finding Slope Given an Equation

When given an equation of a line, it is necessary to solve for y to determine the slope of the line. Given the equation $6x + 2y = 8$, find the slope. First, subtract $6x$ from both sides of the equation, resulting in $2y = -6x + 8$. Then divide both sides of the equation by 2, resulting in $y = -3x + 4$. This then allows us to conclude that the slope of the line is $m = -3$, the coefficient of x. Once an equation is in the form $y = mx + b$, the slope and y-intercept can easily be determined. For this reason, we refer to the equation $y = mx + b$ as "slope-intercept form" of the equation of a line.

Review Video: Finding the Slope of a Line
Visit mometrix.com/academy and enter code: 766664

Linear Equations

Equations like $5x = 100$ and $8x - 120 = 200$ and $6x + 4y = 240$ are **linear equations**. Linear equations are named based off the number of distinct variables they include. For example, the equation $3x + 30 = 8x$ is a **one-variable linear equation** because it involves only the single variable x. It does not matter that x appears more than once. Any equations that can be written as $ax + b = 0$, where $a \neq 0$, falls into this category. Furthermore, the equation $3x - 5y = 14 + 9y$ is a **two-variable linear equation** because it involves the two variables x and y. The equation $7x + 8y - 12z + 14w = 56$ is a linear equation in four variables.

Satisfying the Equation

When given a one-variable linear equation, the goal is typically to solve it. This means that we want to find the number that makes the equation true if we substitute it for the variable. That number is the **solution,** or root, of the equation. For instance, the equation $5x = 10$ has the solution $x = 2$. This is true because when 2 is substituted for x, the result is $5 \cdot 2 = 10$, which is true. On the other hand, $x = 6$ can not be a solution because $5 \cdot 6 \neq 10$, so it is false. Two equations with the same solution are **equivalent equations**. For example, the equations $5x = 10$ and $5x + 3 = 13$ are equivalent because both have the same solution of $x = 2$.

Determining a Solution Set

The **solution set** is the set of all solutions of an equation. In the previous example, the solution set would be 2. Solutions to a linear equation in two variables consist of pairs of numbers. For instance, the equation $6x + 4y = 240$ has the solution $x = 20$ and $y = 30$ since $6 \cdot 20 + 4 \cdot 30 = 240$ is true. We can write this solution as the ordered pair (20,30) and plot it as a point on the coordinate plane. Such equations usually have infinitely many solutions; and if we plot the points for all these solutions we get a line, which is a picture of all the solutions. We call this **graphing the equation**. When an equation has no true solutions, it is referred to as an **empty set**.

Linear Equation Forms

Linear equations can be written many ways. Below is a list of some forms linear equations can take:

- **Standard Form**: $Ax + By = C$; the slope is $\frac{-A}{B}$ and the y-intercept is $\frac{C}{B}$
- **Slope Intercept Form**: $y = mx + b$, where m is the slope and b is the y-intercept
- **Point-Slope Form**: $y - y_1 = m(x - x_1)$, where m is the slope and (x_1, y_1) is a point on the line

- **Two-Point Form**: $\frac{y-y_1}{x-x_1} = \frac{y_2-y_1}{x_2-x_1}$, where (x_1, y_1) and (x_2, y_2) are two points on the given line
- **Intercept Form**: $\frac{x}{x_1} + \frac{y}{y_1} = 1$, where $(x_1, 0)$ is the point at which a line intersects the x-axis, and $(0, y_1)$ is the point at which the same line intersects the y-axis

Review Video: Slope-Intercept and Point-Slope Forms
Visit mometrix.com/academy and enter code: 113216

Review Video: Converting Between Standard and Slope-Intercept Forms
Visit mometrix.com/academy and enter code: 982828

Review Video: Linear Equations Basics
Visit mometrix.com/academy and enter code: 793005

Solving Equations

Manipulating Equations

Like Terms

Like terms are terms in an equation that have the same variable, regardless of whether they also have the same coefficient. This includes terms that *lack* a variable; all constants (i.e., numbers without variables) are considered like terms. If the equation involves terms with a variable raised to different powers, the like terms are those that have the variable raised to the same power.

For example, consider the equation $x^2 + 3x + 2 = 2x^2 + x - 7 + 2x$. In this equation, 2 and –7 are like terms; they are both constants. The terms $3x$, x, and $2x$ are like terms, they all include the variable x raised to the first power. The terms x^2 and $2x^2$ are like terms, they both include the variable x, raised to the second power. The terms $2x$ and $2x^2$ are not like terms; although they both involve the variable x, the variable is not raised to the same power in both terms. The fact that they have the same coefficient, 2, is not relevant.

Review Video: Rules for Manipulating Equations
Visit mometrix.com/academy and enter code: 838871

Carrying Out the Same Operation on Both Sides of an Equation

When solving an equation, the general procedure is to carry out a series of operations on both sides of an equation, choosing operations that simplify the equation when doing so. The reason why the same operation must be carried out on both sides of the equation is because that leaves the meaning of the equation unchanged, and yields a result that is equivalent to the original equation. This would not be the case if we carried out an operation on one side of an equation and not the other. Consider what an equation means: it is a statement that two values or expressions are equal. If we carry out the same operation on both sides of the equation—add 3 to both sides, for example—then the two sides of the equation are changed in the same way, and so remain equal. If we do that to only one side of the equation—add 3 to one side but not the other—then that wouldn't be true; if we change one side of the equation but not the other then the two sides are no longer equal.

Combining Like Terms

Combining like terms refers to adding or subtracting like terms—terms with the same variable—and therefore reducing sets of like terms to a single term. The main advantage of doing this is that it simplifies the equation. Often, combining like terms can be done as the first step in solving an equation, though it can also be done later, such as after distributing terms in a product.

For example, consider the equation $2(x+3)+3(2+\ x+3)=-4$. The 2 and the 3 in the second set of parentheses are like terms, and we can combine them, yielding $2(x+3)+3(x+5)=-4$. Now we can carry out the multiplications implied by the parentheses, distributing the outer 2 and 3 accordingly: $2x+6+3x+15=-4$. The $2x$ and the $3x$ are like terms, and we can add them together: $5x+6+15=-4$. Now, the constants 6, 15, and –4 are also like terms, and we can combine them as well: subtracting 6 and 15 from both sides of the equation, we get $5x=-4-6-15$, or $5x=-25$, which simplifies further to $x=-5$.

Review Video: Solving Equations by Combining Like Terms
Visit mometrix.com/academy and enter code: 668506

Canceling Terms on Opposite Sides of an Equation

Two terms on opposite sides of an equation can be canceled if and only if they *exactly* match each other. They must have the same variable raised to the same power and the same coefficient. For example, in the equation $3x+2x^2+6=2x^2-6$, $2x^2$ appears on both sides of the equation and can be canceled, leaving $3x+6=-6$. The 6 on each side of the equation *cannot* be canceled, because it is added on one side of the equation and subtracted on the other. While they cannot be canceled, however, the 6 and –6 are like terms and can be combined, yielding $3x=-12$, which simplifies further to $x=-4$.

It's also important to note that the terms to be canceled must be independent terms and cannot be part of a larger term. For example, consider the equation $2(x+6)=3(x+4)+1$. We cannot cancel the x's, because even though they match each other they are part of the larger terms $2(x+6)$ and $3(x+4)$. We must first distribute the 2 and 3, yielding $2x+12=3x+12+1$. Now we see that the terms with the x's do not match, but the 12s do, and can be canceled, leaving $2x=3x+1$, which simplifies to $x=-1$.

Isolating Variables

To isolate a variable means to manipulate the equation so that the variable appears by itself on one side of the equation, and does not appear at all on the other side. Generally, an equation or inequality is considered to be solved once the variable is isolated and the other side of the equation or inequality is simplified as much as possible. In the case of a two-variable equation or inequality, only one variable needs to be isolated; it will not usually be possible to simultaneously isolate both variables.

For a linear equation—an equation in which the variable only appears raised to the first power—isolating a variable can be done by first moving all the terms with the variable to one side of the equation and all other terms to the other side. (*Moving* a term really means adding the inverse of the term to both sides; when a term is *moved* to the other side of the equation its sign is flipped.) Then combine like terms on each side. Finally, divide both sides by the coefficient of the variable, if applicable. The steps need not necessarily be done in this order, but this order will always work.

Review Video: Solving Equations for Specific Variables
Visit mometrix.com/academy and enter code: 130695

Review Video: Solving Equations Involving Algebraic Fractions
Visit mometrix.com/academy and enter code: 237770

Review Video: Solving One-Step Equations
Visit mometrix.com/academy and enter code: 777004

Solving One-Variable Linear Equations

Equations with One Solution (the usual case)

To solve a one-variable linear equation, we use the techniques above to isolate the variable.

1. If any coefficients or constants are fractions, it is often helpful first to multiply both sides of the equation by the least common denominator (of all fractions) to clear the fractions.
2. Simplify both sides of the equation by combining any like terms.
3. Put all terms with the variable on one side of the equation and all constant terms on the other side, by adding or subtracting the same terms on both sides of the equation.
4. Divide both sides by the coefficient of the variable (or multiply both sides by its reciprocal).
5. When we have a value for the variable, we can check it by substituting the value into the original equation to make sure it produces a true result.

Consider the following example for solving the equation $\frac{2}{3}x + 8 = 14$:

$3 \cdot \left(\frac{2}{3}x + 8\right) = 3 \cdot 14$	Clear fractions by multiplying both sides by 3.
$2x + 24 = 42$	Simplify, remembering to apply the distributive property.
$2x + 24 - 24 = 42 - 24$	Subtract 24 from both sides to isolate $2x$.
$2x = 18$	Simplify by combining like terms.
$\frac{2x}{2} = \frac{18}{2}$	Divide both sides by 2 to isolate x.
$x = 9$	Simplify

Finally, we check this answer by substituting $x = 9$ into the original equation to make sure we get a true result.

$$\frac{2}{3}x + 8 = \frac{2}{3}(9) + 8 = 6 + 8 = 14$$

This is correct, so the value of x is 9.

Review Video: Solving Equations Using the Distributive Property
Visit mometrix.com/academy and enter code: 765499

Equations with More Than One Solution

Some types of non-linear equations, such as equations involving squares of variables, may have more than one solution. For example, the equation $x^2 = 4$ has two solutions: 2 and –2. Equations with absolute values can also have multiple solutions: $|x| = 1$ has the solutions $x = 1$ and $x = -1$.

It is possible for a linear equation to have more than one solution but only if the equation is true regardless of the value of the variable. We call such an equation an **identity**. In this case, the equation has infinitely many solutions, because every possible value of the variable is a solution. We discover that a linear equation is an identity when our attempts to isolate the variable cause the variable to disappear, leaving a *true* equation involving only constants. For example, consider the equation $2(3x + 5) = x + 5(x + 2)$. Distributing, we get $6x + 10 = x + 5x + 10$; combining like terms gives $6x + 10 = 6x + 10$, and the $6x$-terms cancel to leave $10 = 10$. This is clearly true, so the original equation is an identity. We could also cancel the 10's leaving $0 = 0$, which is also is clearly true—in general if both sides of the equation can be reduced to match one another exactly, the original equation is an identity.

Equations with No Solution

Some types of non-linear equations, such as equations involving squares of variables, may have no solution. For example, the equation $x^2 = -2$ has no solutions in the real numbers because the square of a real number must be positive. Similarly, $|x| = -1$ has no solution because the absolute value of a number is always positive.

It is also possible for a linear equation to have no solution. We call such an equation a **contradiction**. We discover that a linear equation is a contradiction when our attempts to isolate the variable cause the variable to disappear, leaving a *false* equation involving only constants. For example, the equation $2(x + 3) + x = 3x$ has no solution. We can see this by trying to solve it: first we distribute, leaving $2x + 6 + x = 3x$. Combining like terms gives us $3x + 6 = 3x$, and cancelling the term $3x$ on both sides leaves us with $6 = 0$. This is clearly false, so the original equation is a contradiction, having no solutions.

Features of Equations That Require Special Treatment

A linear equation is an equation in which variables only appear by themselves: not multiplied together, not with exponents other than one, and not inside absolute value signs or any other functions. For example, the equation $x + 1 - 3x = 5 - x$ is a linear equation; while x appears multiple times, it never appears with an exponent other than one, or inside any function. The two-variable equation $2x - 3y = 5 + 2x$ is also a linear equation. In contrast, the equation $x^2 - 5 = 3x$ is *not* a linear equation, because it involves the term x^2. The equation $\sqrt{x} = 5$ is not linear, because it involves a square root. The equation $(x - 1)^2 = 4$ is not linear because even though there's no exponent on the x directly, it appears as part of an expression that is squared. The two-variable equation $x + xy - y = 5$ is not linear because it includes the term xy, where two variables are multiplied together.

As we see above, linear equations can always be solved (or shown to have no solution) by combining like terms and performing simple operations on both sides of the equation. Some non-linear equations can be solved by similar methods, but others may require more advanced methods of solution, if they can be solved analytically at all.

Solving Equations Involving Roots

In an equation involving roots, the first step is to isolate the term with the root, if possible, and then raise both sides of the equation to the appropriate power to eliminate it. Consider an example equation, $2\sqrt{x + 1} - 1 = 3$. In this case, begin by adding 1 to both sides, yielding $2\sqrt{x + 1} = 4$, and then dividing both sides by 2, yielding $\sqrt{x + 1} = 2$. Now square both sides, yielding $x + 1 = 4$. Finally, subtracting 1 from both sides yields $x = 3$.

Squaring both sides of an equation (or raising both sides to any *even* power) may, however, yield a spurious solution—a solution to the squared equation that is *not* a solution of the original equation. It's therefore necessary to plug the solution back into the original equation to make sure it works. In this case, it does: $2\sqrt{3 + 1} - 1 = 2\sqrt{4} - 1 = 2(2) - 1 = 4 - 1 = 3$.

The same procedure applies for other roots as well. For example, given the equation $3 + \sqrt[3]{2x} = 5$, we can first subtract 3 from both sides, yielding $\sqrt[3]{2x} = 2$ and isolating the root. Raising both sides to the third power yields $2x = 2^3$; i.e., $2x = 8$. We can now divide both sides by 2 to get $x = 4$.

Review Video: Solving Equations Involving Roots
Visit mometrix.com/academy and enter code: 297670

Solving Equations with Exponents

In solving an equation with powers of a variable, sometimes it is possible to eliminate all but one term involving the variable. In that case, we can isolate the power of the variable and then take the appropriate root of both sides to eliminate the exponent. For instance, for the equation $2x^3 + 17 = 5x^3 - 7$, we can subtract $5x^3$ from both sides to get $-3x^3 + 17 = -7$, and then subtract 17 from both sides to get $-3x^3 = -24$. Finally, we can divide both sides by -3 to get $x^3 = 8$. Since this isolates the cube of the variable, we can take the cube root of both sides to get $x = \sqrt[3]{8} = 2$.

One important but often overlooked point is that equations with an exponent greater than 1 may have more than one answer. The solution to $x^2 = 9$ isn't simply $x = 3$; it's $x = \pm 3$ (that is, $x = 3$ or $x = -3$). For a slightly more complicated example, consider the equation $(x - 1)^2 - 1 = 3$. Adding 1 to both sides yields $(x - 1)^2 = 4$; taking the square root of both sides yields $x - 1 = 2$. We can then add 1 to both sides to get $x = 3$. However, there's a second solution. We also have the possibility that $x - 1 = -2$, in which case $x = -1$. Both $x = 3$ and $x = -1$ are valid solutions, as can be verified by substituting them both into the original equation.

Review Video: Solving Equations with Exponents
Visit mometrix.com/academy and enter code: 514557

Review Video: Adding and Subtracting with Exponents
Visit mometrix.com/academy and enter code: 875756

Solving Equations with Absolute Values

When solving an equation with an absolute value, the first step is to isolate the absolute value term. We then consider two possibilities: when the expression inside the absolute value is positive or when it is negative. In the former case, the expression in the absolute value equals the expression on the other side of the equation; in the latter, it equals the additive inverse of that expression—the expression times negative one. We consider each case separately and finally check for spurious solutions.

For instance, consider solving $|2x - 1| + x = 5$ for x. We can first isolate the absolute value by moving the x to the other side: $|2x - 1| = -x + 5$. Now, we have two possibilities. First, that $2x - 1$ is positive, and hence $2x - 1 = -x + 5$. Rearranging and combining like terms yields $3x = 6$, and hence $x = 2$. The other possibility is that $2x - 1$ is negative, and hence $2x - 1 = -(-x + 5) = x - 5$. In this case, rearranging and combining like terms yields $x = -4$. Substituting $x = 2$ and $x = -4$ back into the original equation, we see that they are both valid solutions.

Note that the absolute value of a sum or difference applies to the sum or difference as a whole, not to the individual terms; in general, $|2x - 1|$ is not equal to $|2x + 1|$ or to $|2x| - 1$.

Review Video: Solving Absolute Value Equations
Visit mometrix.com/academy and enter code: 501208

EXTRANEOUS SOLUTIONS

An **extraneous solution** may arise when we square both sides of an equation (or raise both sides to an even power) as a step in solving it or under certain other operations on the equation. It is a solution to the squared or otherwise modified equation that is *not* a solution of the original equation. To identify an extraneous solution, it's useful when you solve an equation involving roots or absolute values to plug the solution back into the original equation to make sure it's valid.

TWO-VARIABLE EQUATIONS

Similar to methods for a one-variable equation, solving a two-variable equation involves isolating a variable: manipulating the equation so that a variable appears by itself on one side of the equation, and not at all on the other side. However, in a two-variable equation, you will usually only be able to isolate one of the variables; the other variable may appear on the other side along with constant terms, or with exponents or other functions. If an equation has multiple variables, the problem should tell you which variable to isolate.

Review Video: Solving Equations with Variables on Both Sides
Visit mometrix.com/academy and enter code: 402497

GRAPHING EQUATIONS

GRAPHICAL SOLUTIONS TO EQUATIONS

When equations are shown graphically, they are usually shown on a **Cartesian coordinate plane**. The Cartesian coordinate plane consists of two number lines placed perpendicular to each other and intersecting at the zero point, also known as the origin. The horizontal number line is known as the x-axis, with positive values to the right of the origin, and negative values to the left of the origin. The vertical number line is known as the y-axis, with positive values above the origin, and negative values below the origin. Any point on the plane can be identified by an ordered pair in the form (x, y), called coordinates. The x-value of the coordinate is called the abscissa, and the y-value of the coordinate is called the ordinate. The two number lines divide the plane into **four quadrants**: I, II, III, and IV.

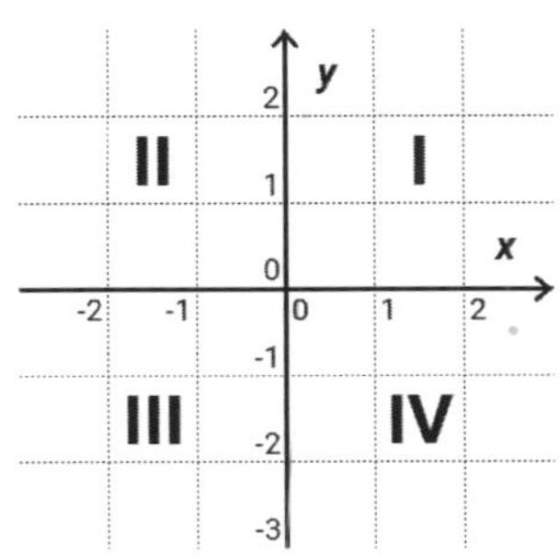

Note that in quadrant I $x > 0$ and $y > 0$, in quadrant II $x < 0$ and $y > 0$, in quadrant III $x < 0$ and $y < 0$, and in quadrant IV $x > 0$ and $y < 0$.

Recall that if the value of the slope of a line is positive, the line slopes upward from left to right. If the value of the slope is negative, the line slopes downward from left to right. If the y-coordinates are the same for two points on a line, the slope is 0 and the line is a **horizontal line**. If the x-coordinates are the same for two points on a line, there is no slope and the line is a **vertical line**.

Two or more lines that have equivalent slopes are **parallel lines**. **Perpendicular lines** have slopes that are negative reciprocals of each other, such as $\frac{a}{b}$ and $\frac{-b}{a}$.

> **Review Video: Cartesian Coordinate Plane and Graphing**
> Visit mometrix.com/academy and enter code: 115173

Graphing Equations in Two Variables

One way of graphing an equation in two variables is to plot enough points to get an idea for its shape and then draw the appropriate curve through those points. A point can be plotted by substituting in a value for one variable and solving for the other. If the equation is linear, we only need two points and can then draw a straight line between them.

For example, consider the equation $y = 2x - 1$. This is a linear equation—both variables only appear raised to the first power—so we only need two points. When $x = 0$, $y = 2(0) - 1 = -1$. When $x = 2$, $y = 2(2) - 1 = 3$. We can therefore choose the points $(0, -1)$ and $(2, 3)$, and draw a line between them:

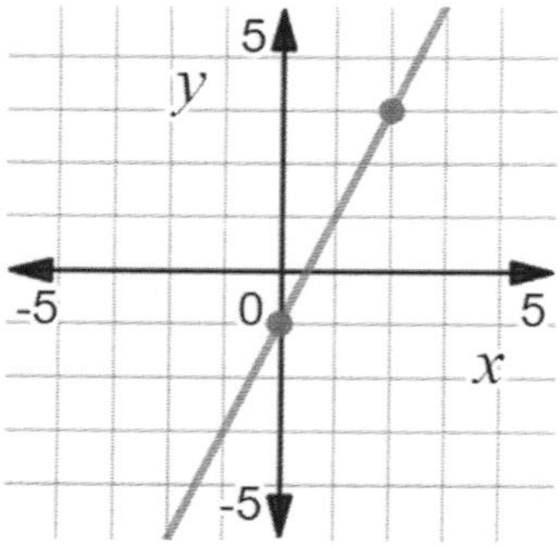

Inequalities

Inequalities

Commonly in algebra and other upper-level fields of math you find yourself working with mathematical expressions that do not equal each other. The statement comparing such expressions with symbols such as $<$ (less than) or $>$ (greater than) is called an *inequality*. An example of an inequality is $7x > 5$. To solve for x, simply divide both sides by 7 and the solution is shown to be $x > \frac{5}{7}$. Graphs of the solution set of inequalities are represented on a number line. Open circles are used to show that an expression approaches a number but is never quite equal to that number.

> **Review Video: Solving One-Step Inequalities**
> Visit mometrix.com/academy and enter code: 229684
>
> **Review Video: Solving Multi-Step Inequalities**
> Visit mometrix.com/academy and enter code: 347842
>
> **Review Video: Solving Inequalities Using All 4 Basic Operations**
> Visit mometrix.com/academy and enter code: 401111

TYPES OF INEQUALITIES

Conditional inequalities are those with certain values for the variable that will make the condition true and other values for the variable where the condition will be false. **Absolute inequalities** can have any real number as the value for the variable to make the condition true, while there is no real number value for the variable that will make the condition false. Solving inequalities is done by following the same rules for solving equations with the exception that when multiplying or dividing by a negative number the direction of the inequality sign must be flipped or reversed. **Double inequalities** are situations where two inequality statements apply to the same variable expression. Example: $-c < ax + b < c$.

> **Review Video: Conditional and Absolute Inequalities**
> Visit mometrix.com/academy and enter code: 980164

SOLVING INEQUALITIES

DETERMINING SOLUTIONS TO INEQUALITIES

To determine whether a coordinate is a solution of an inequality, you can substitute the values of the coordinate into the inequality, simplify, and check whether the resulting statement holds true. For instance, to determine whether $(-2,4)$ is a solution of the inequality $y \geq -2x + 3$, substitute the values into the inequality, $4 \geq -2(-2) + 3$. Simplify the right side of the inequality and the result is $4 \geq 7$, which is a false statement. Therefore, the coordinate is not a solution of the inequality. You can also use this method to determine which part of the graph of an inequality is shaded. The graph of $y \geq -2x + 3$ includes the solid line $y = -2x + 3$ and, since it excludes the point $(-2,4)$ to the left of the line, it is shaded to the right of the line.

> **Review Video: Graphing Linear Inequalities**
> Visit mometrix.com/academy and enter code: 439421
>
> **Review Video: Graphing Solutions to Inequalities**
> Visit mometrix.com/academy and enter code: 391281

FLIPPING INEQUALITY SIGNS

When given an inequality, we can always turn the entire inequality around, swapping the two sides of the inequality and changing the inequality sign. For instance, $x + 2 > 2x - 3$ is equivalent to $2x - 3 < x + 2$. Aside from that, normally the inequality does not change if we carry out the same operation on both sides of the inequality. There is, however, one principal exception: if we *multiply* or *divide* both sides of the inequality by a *negative number*, the inequality is flipped. For example, if we take the inequality $-2x < 6$ and divide both sides by −2, the inequality flips and we are left with $x > -3$. This *only* applies to multiplication and division, and only with negative numbers. Multiplying or dividing both sides by a positive number, or adding or subtracting any number regardless of sign, does not flip the inequality. Another special case that flips the inequality sign is when reciprocals are used. For instance, $3 > 2$ but the relation of the reciprocals is $\frac{1}{3} < \frac{1}{2}$.

COMPOUND INEQUALITIES

A **compound inequality** is an equality that consists of two inequalities combined with *and* or *or*. The two components of a proper compound inequality must be of opposite type: that is, one must be greater than (or greater than or equal to), the other less than (or less than or equal to). For instance, "$x + 1 < 2$ or $x + 1 > 3$" is a compound inequality, as is "$2x \geq 4$ and $2x \leq 6$." An *and* inequality can be written more compactly by having one inequality on each side of the common part: "$2x \geq 1$ and $2x \leq 6$," can also be written as $1 \leq 2x \leq 6$.

In order for the compound inequality to be meaningful, the two parts of an *and* inequality must overlap; otherwise, no numbers satisfy the inequality. On the other hand, if the two parts of an *or* inequality overlap, then *all* numbers satisfy the inequality and as such the inequality is usually not meaningful.

Solving a compound inequality requires solving each part separately. For example, given the compound inequality "$x + 1 < 2$ or $x + 1 > 3$," the first inequality, $x + 1 < 2$, reduces to $x < 1$, and the second part, $x + 1 > 3$, reduces to $x > 2$, so the whole compound inequality can be written as "$x < 1$ or $x > 2$." Similarly, $1 \leq 2x \leq 6$ can be solved by dividing each term by 2, yielding $\frac{1}{2} \leq x \leq 3$.

Review Video: Compound Inequalities
Visit mometrix.com/academy and enter code: 786318

Solving Inequalities Involving Absolute Values

To solve an inequality involving an absolute value, first isolate the term with the absolute value. Then proceed to treat the two cases separately as with an absolute value equation, but flipping the inequality in the case where the expression in the absolute value is negative (since that essentially involves multiplying both sides by −1.) The two cases are then combined into a compound inequality; if the absolute value is on the greater side of the inequality, then it is an *or* compound inequality, if on the lesser side, then it's an *and*.

Consider the inequality $2 + |x - 1| \geq 3$. We can isolate the absolute value term by subtracting 2 from both sides: $|x - 1| \geq 1$. Now, we're left with the two cases $x - 1 \geq 1$ or $x - 1 \leq -1$: note that in the latter, negative case, the inequality is flipped. $x - 1 \geq 1$ reduces to $x \geq 2$, and $x - 1 \leq -1$ reduces to $x \leq 0$. Since in the inequality $|x - 1| \geq 1$ the absolute value is on the greater side, the two cases combine into an *or* compound inequality, so the final, solved inequality is "$x \leq 0$ or $x \geq 2$."

Review Video: Solving Absolute Value Inequalities
Visit mometrix.com/academy and enter code: 997008

Solving Inequalities Involving Square Roots

Solving an inequality with a square root involves two parts. First, we solve the inequality as if it were an equation, isolating the square root and then squaring both sides of the equation. Second, we restrict the solution to the set of values of x for which the value inside the square root sign is non-negative.

For example, in the inequality, $\sqrt{x - 2} + 1 < 5$, we can isolate the square root by subtracting 1 from both sides, yielding $\sqrt{x - 2} < 4$. Squaring both sides of the inequality yields $x - 2 < 16$, so $x < 18$. Since we can't take the square root of a negative number, we also require the part inside the square root to be non-negative. In this case, that means $x - 2 \geq 0$. Adding 2 to both sides of the inequality yields $x \geq 2$. Our final answer is a compound inequality combining the two simple inequalities: $x \geq 2$ and $x < 18$, or $2 \leq x < 18$.

Note that we only get a compound inequality if the two simple inequalities are in opposite directions; otherwise, we take the one that is more restrictive.

The same technique can be used for other even roots, such as fourth roots. It is *not*, however, used for cube roots or other odd roots—negative numbers *do* have cube roots, so the condition that the quantity inside the root sign cannot be negative does not apply.

> **Review Video: Solving Inequalities Involving Square Roots**
> Visit mometrix.com/academy and enter code: 800288

SPECIAL CIRCUMSTANCES

Sometimes an inequality involving an absolute value or an even exponent is true for all values of x, and we don't need to do any further work to solve it. This is true if the inequality, once the absolute value or exponent term is isolated, says that term is greater than a negative number (or greater than or equal to zero). Since an absolute value or a number raised to an even exponent is *always* non-negative, this inequality is always true.

GRAPHING INEQUALITIES

GRAPHING SIMPLE INEQUALITIES

To graph a simple inequality, we first mark on the number line the value that signifies the end point of the inequality. If the inequality is strict (involves a less than or greater than), we use a hollow circle; if it is not strict (less than or equal to or greater than or equal to), we use a solid circle. We then fill in the part of the number line that satisfies the inequality: to the left of the marked point for less than (or less than or equal to), to the right for greater than (or greater than or equal to).

For example, we would graph the inequality $x < 5$ by putting a hollow circle at 5 and filling in the part of the line to the left:

GRAPHING COMPOUND INEQUALITIES

To graph a compound inequality, we fill in both parts of the inequality for an *or* inequality, or the overlap between them for an *and* inequality. More specifically, we start by plotting the endpoints of each inequality on the number line. For an *or* inequality, we then fill in the appropriate side of the line for each inequality. Typically, the two component inequalities do not overlap, which means the shaded part is *outside* the two points. For an *and* inequality, we instead fill in the part of the line that meets both inequalities.

For the inequality "$x \leq -3$ or $x > 4$," we first put a solid circle at –3 and a hollow circle at 4. We then fill the parts of the line *outside* these circles:

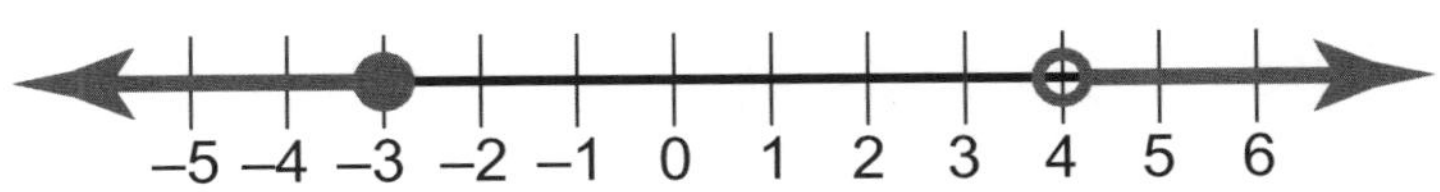

GRAPHING INEQUALITIES INCLUDING ABSOLUTE VALUES

An inequality with an absolute value can be converted to a compound inequality. To graph the inequality, first convert it to a compound inequality, and then graph that normally. If the absolute value is on the greater side of the inequality, we end up with an *or* inequality; we plot the endpoints of the inequality on the number line and fill in the part of the line *outside* those points. If the absolute value is on the smaller side of the inequality, we end up with an *and* inequality; we plot the endpoints of the inequality on the number line and fill in the part of the line *between* those points.

For example, the inequality $|x + 1| \geq 4$ can be rewritten as $x \geq 3$ or $x \leq -5$. We place solid circles at the points 3 and -5 and fill in the part of the line *outside* them:

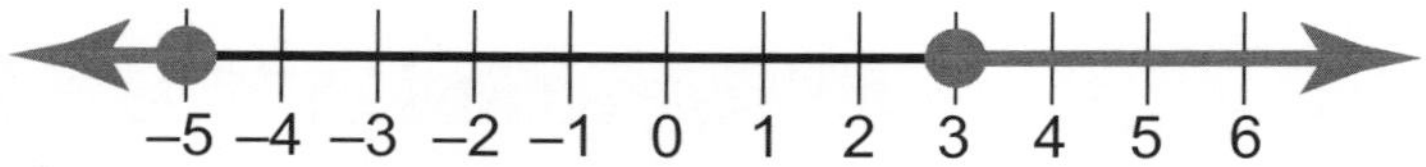

Graphing Inequalities in Two Variables

To graph an inequality in two variables, we first graph the border of the inequality. This means graphing the equation that we get if we replace the inequality sign with an equals sign. If the inequality is strict ($>$ or $<$), we graph the border with a dashed or dotted line; if it is not strict ($\geq$ or $\leq$), we use a solid line. We can then test any point not on the border to see if it satisfies the inequality. If it does, we shade in that side of the border; if not, we shade in the other side. As an example, consider $y > 2x + 2$. To graph this inequality, we first graph the border, $y = 2x + 2$. Since it is a strict inequality, we use a dashed line. Then, we choose a test point. This can be any point not on the border; in this case, we will choose the origin, (0,0). (This makes the calculation easy and is generally a good choice unless the border passes through the origin.) Putting this into the original inequality, we get $0 > 2(0) + 2$, i.e., $0 > 2$. This is *not* true, so we shade in the side of the border that does *not* include the point (0,0):

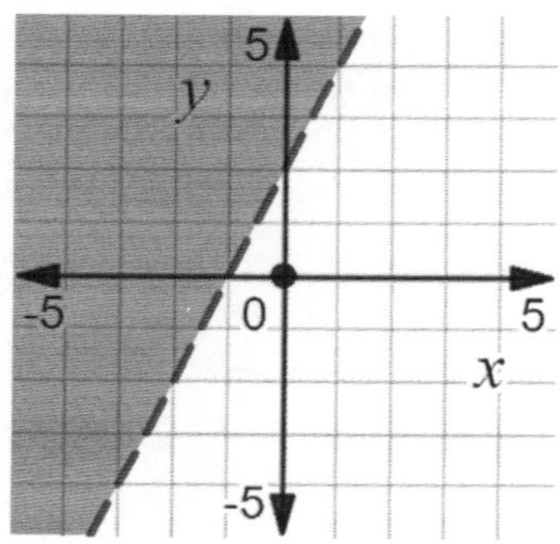

Graphing Compound Inequalities in Two Variables

One way to graph a compound inequality in two variables is to first graph each of the component inequalities. For an *and* inequality, we then shade in only the parts where the two graphs overlap; for an *or* inequality, we shade in any region that pertains to either of the individual inequalities.

Consider the graph of "$y \geq x - 1$ and $y \leq -x$":

We first shade in the individual inequalities:

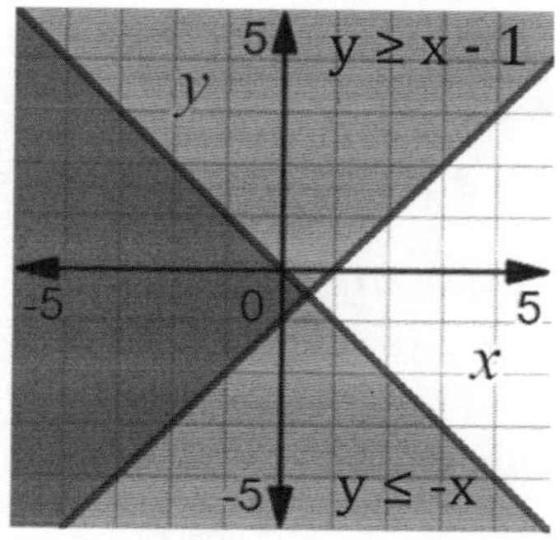

Now, since the compound inequality has an *and*, we only leave shaded the overlap—the part that pertains to *both* inequalities:

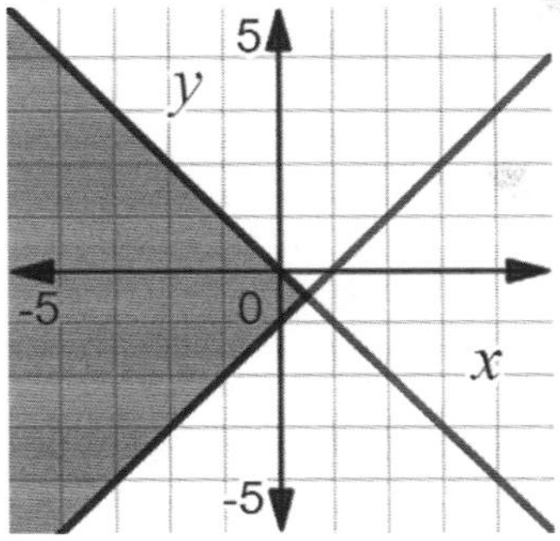

If instead the inequality had been "$y \geq x - 1$ or $y \leq -x$," our final graph would involve the *total* shaded area:

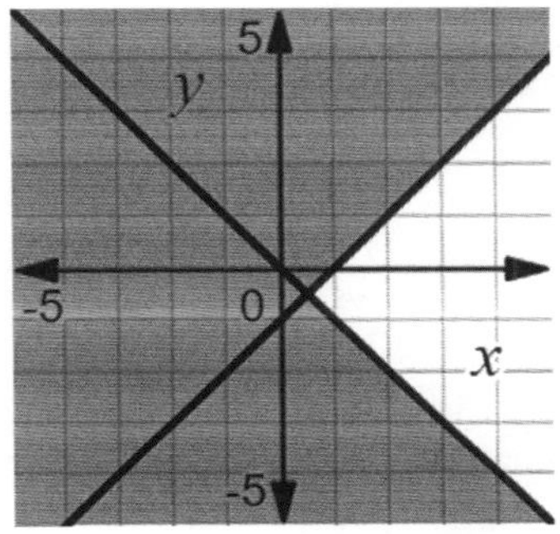

Review Video: Graphing Solutions to Inequalities
Visit mometrix.com/academy and enter code: 391281

Advanced Expressions and Equations

Systems of Equations

Solving Systems of Equations

A **system of equations** is a set of simultaneous equations that all use the same variables. A solution to a system of equations must be true for each equation in the system. **Consistent systems** are those with at least one solution. **Inconsistent systems** are systems of equations that have no solution.

Review Video: Solving Systems of Linear Equations
Visit mometrix.com/academy and enter code: 746745

Substitution

To solve a system of linear equations by **substitution**, start with the easier equation and solve for one of the variables. Express this variable in terms of the other variable. Substitute this expression in the other equation and solve for the other variable. The solution should be expressed in the form

(x, y). Substitute the values into both of the original equations to check your answer. Consider the following system of equations:

$$x + 6y = 15$$
$$3x - 12y = 18$$

Solving the first equation for x: $x = 15 - 6y$

Substitute this value in place of x in the second equation, and solve for y:

$$\begin{aligned} 3(15 - 6y) - 12y &= 18 \\ 45 - 18y - 12y &= 18 \\ 30y &= 27 \\ y &= \frac{27}{30} = \frac{9}{10} = 0.9 \end{aligned}$$

Plug this value for y back into the first equation to solve for x:

$$x = 15 - 6(0.9) = 15 - 5.4 = 9.6$$

Check both equations if you have time:

$$\begin{aligned} 9.6 + 6(0.9) &= 15 \\ 9.6 + 5.4 &= 15 \\ 15 &= 15 \end{aligned} \qquad \begin{aligned} 3(9.6) - 12(0.9) &= 18 \\ 28.8 - 10.8 &= 18 \\ 18 &= 18 \end{aligned}$$

Therefore, the solution is (9.6,0.9).

Review Video: The Substitution Method
Visit mometrix.com/academy and enter code: 565151

Review Video: Substitution and Elimination
Visit mometrix.com/academy and enter code: 958611

ELIMINATION

To solve a system of equations using **elimination**, begin by rewriting both equations in standard form $Ax + By = C$. Check to see if the coefficients of one pair of like variables add to zero. If not, multiply one or both of the equations by a non-zero number to make one set of like variables add to zero. Add the two equations to solve for one of the variables. Substitute this value into one of the original equations to solve for the other variable. Check your work by substituting into the other equation. Now, let's look at solving the following system using the elimination method:

$$5x + 6y = 4$$
$$x + 2y = 4$$

If we multiply the second equation by -3, we can eliminate the y-terms:

$$5x + 6y = 4$$
$$-3x - 6y = -12$$

Add the equations together and solve for x:

$$2x = -8$$
$$x = \frac{-8}{2} = -4$$

Plug the value for x back in to either of the original equations and solve for y:

$$-4 + 2y = 4$$
$$y = \frac{4+4}{2} = 4$$

Check both equations if you have time:

$$5(-4) + 6(4) = 4 \qquad -4 + 2(4) = 4$$
$$-20 + 24 = 4 \qquad -4 + 8 = 4$$
$$4 = 4 \qquad 4 = 4$$

Therefore, the solution is $(-4,4)$.

Review Video: The Elimination Method
Visit mometrix.com/academy and enter code: 449121

GRAPHICALLY

To solve a system of linear equations **graphically**, plot both equations on the same graph. The solution of the equations is the point where both lines cross. If the lines do not cross (are parallel), then there is **no solution**.

For example, consider the following system of equations:

$$y = 2x + 7$$
$$y = -x + 1$$

Since these equations are given in slope-intercept form, they are easy to graph; the y-intercepts of the lines are $(0,7)$ and $(0,1)$. The respective slopes are 2 and –1, thus the graphs look like this:

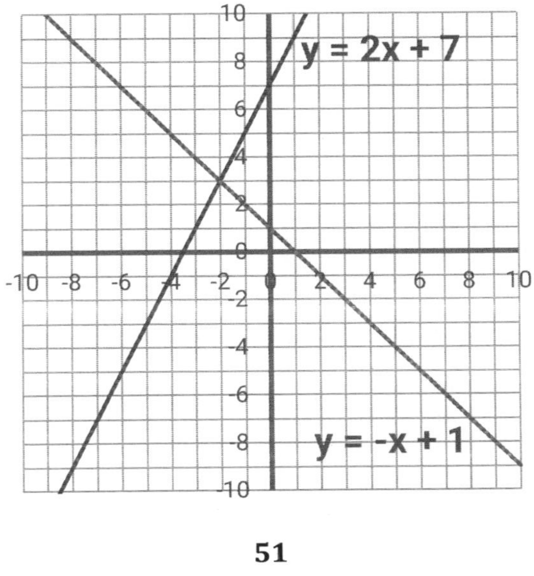

The two lines intersect at the point $(-2,3)$, thus this is the solution to the system of equations.

Solving a system graphically is generally only practical if both coordinates of the solution are integers; otherwise the intersection will lie between gridlines on the graph and the coordinates will be difficult or impossible to determine exactly. It also helps if, as in this example, the equations are in slope-intercept form or some other form that makes them easy to graph. Otherwise, another method of solution (by substitution or elimination) is likely to be more useful.

Review Video: Solving Systems by Graphing
Visit mometrix.com/academy and enter code: 634812

Solving Systems of Equations Using the Trace Feature

Using the trace feature on a calculator requires that you rewrite each equation, isolating the y-variable on one side of the equal sign. Enter both equations in the graphing calculator and plot the graphs simultaneously. Use the trace cursor to find where the two lines cross. Use the zoom feature if necessary to obtain more accurate results. Always check your answer by substituting into the original equations. The trace method is likely to be less accurate than other methods due to the resolution of graphing calculators but is a useful tool to provide an approximate answer.

Advanced Systems of Equations

Solving a System of Equations with a Linear Equation and a Quadratic Equation

Algebraically

Generally, the simplest way to solve a system of equations consisting of a linear equation and a quadratic equation algebraically is through the method of substitution. One possible strategy is to solve the linear equation for y and then substitute that expression into the quadratic equation. After expansion and combining like terms, this will result in a new quadratic equation for x, which, like all quadratic equations, may have zero, one, or two solutions. Plugging each solution for x back into one of the original equations will then produce the corresponding value of y.

For example, consider the following system of equations:

$$x + y = 1$$
$$y = (x + 3)^2 - 2$$

We can solve the linear equation for y to yield $y = -x + 1$. Substituting this expression into the quadratic equation produces $-x + 1 = (x + 3)^2 - 2$. We can simplify this equation:

$$-x + 1 = (x + 3)^2 - 2$$
$$-x + 1 = x^2 + 6x + 9 - 2$$
$$-x + 1 = x^2 + 6x + 7$$
$$0 = x^2 + 7x + 6$$

This quadratic equation can be factored as $(x + 1)(x + 6) = 0$. It therefore has two solutions: $x_1 = -1$ and $x_2 = -6$. Plugging each of these back into the original linear equation yields $y_1 = -x_1 + 1 = -(-1) + 1 = 2$ and $y_2 = -x_2 + 1 = -(-6) + 1 = 7$. Thus, this system of equations has two solutions, $(-1,2)$ and $(-6,7)$.

It may help to check your work by putting each x- and y-value back into the original equations and verifying that they do provide a solution.

Graphically

To solve a system of equations consisting of a linear equation and a quadratic equation graphically, plot both equations on the same graph. The linear equation will, of course, produce a straight line, while the quadratic equation will produce a parabola. These two graphs will intersect at zero, one, or two points; each point of intersection is a solution of the system.

For example, consider the following system of equations:

$$y = -2x + 2$$
$$y = -2x^2 + 4x + 2$$

The linear equation describes a line with a y-intercept of $(0,2)$ and a slope of -2.

To graph the quadratic equation, we can first find the vertex of the parabola: the x-coordinate of the vertex is $h = -\frac{b}{2a} = -\frac{4}{2(-2)} = 1$, and the y-coordinate is $k = -2(1)^2 + 4(1) + 2 = 4$. Thus, the vertex lies at $(1,4)$. To get a feel for the rest of the parabola, we can plug in a few more values of x to find more points; by putting in $x = 2$ and $x = 3$ in the quadratic equation, we find that the points $(2,2)$ and $(3,-4)$ lie on the parabola; by symmetry, so must $(0,2)$ and $(-1,-4)$. We can now plot both equations:

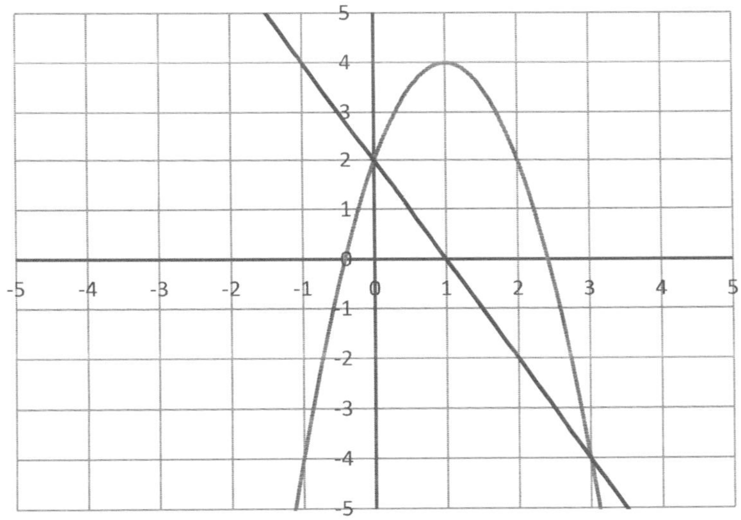

These two curves intersect at the points $(0,2)$ and $(3,-4)$, thus these are the solutions of the equation.

Review Video: Parabolas
Visit mometrix.com/academy and enter code: 129187

Review Video: Vertex of a Parabola
Visit mometrix.com/academy and enter code: 272300

Review Video: Solving a System of Linear and Quadratic Equations
Visit mometrix.com/academy and enter code: 194870

Midpoint and Distance Formulas

If you know the coordinates of the endpoints of a line segment, you can calculate the midpoint and length of the line segment. Conveniently, the length of the line segment is also the distance between the two endpoints.

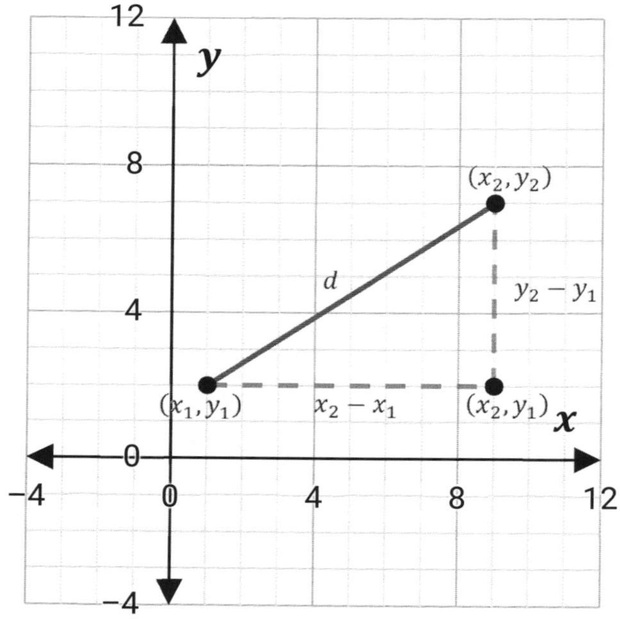

To find the **midpoint** of the line segment with endpoints (x_1, y_1) and (x_2, y_2), average the x-coordinates to get the x-coordinate of the midpoint, and average the y-coordinates to get the y-coordinate of the midpoint. Thus, the **midpoint formula** is:

$$\left(\frac{x_1 + x_2}{2}, \frac{y_1 + y_2}{2}\right)$$

The **distance** between points (x_1, y_1) and (x_2, y_2) is the same as the length of the hypotenuse of a right triangle with the two given points as endpoints, and the two sides of the right triangle parallel to the x-axis and y-axis, respectively. The length of the segment parallel to the x-axis is the difference between the x-coordinates of the two points. The length of the segment parallel to the y-axis is the difference between the y-coordinates of the two points. Use the Pythagorean theorem $a^2 + b^2 = c^2$ or $c = \sqrt{a^2 + b^2}$ to find the distance. Thus, the **distance formula** is:

$$d = \sqrt{(x_2 - x_1)^2 + (y_2 - y_1)^2}$$

Review Video: Calculations Using Points on a Graph
Visit mometrix.com/academy and enter code: 883228

POLYNOMIALS

MONOMIALS AND POLYNOMIALS

A **monomial** is a single constant, variable, or product of constants and variables, such as 7, x, $2x$, or x^3y. There will never be addition or subtraction symbols in a monomial. Like monomials have like variables, but they may have different coefficients. A **polynomial** is a monomial or the result of combining two or more monomials by sums or differences. In a polynomial we call each monomial a **term**. Two terms make a **binomial** (e.g., $2x + 3y$), three terms make a **trinomial** (e.g., $5x^2 - 4x + 9$). The **degree of a monomial** is the sum of the exponents of the variables. The **degree of a polynomial** is the highest degree of any individual term.

Review Video: Polynomials
Visit mometrix.com/academy and enter code: 305005

SIMPLIFYING POLYNOMIALS

Simplifying polynomials requires combining like terms. The like terms in a polynomial expression are those that have the same variables raised to the same powers. It is often helpful to connect the like terms with arrows or lines in order to separate them from the other monomials. Once you have determined the like terms, you can rearrange the polynomial by placing them together. Remember to include the sign that is in front of each term. Once the like terms are placed together, you can apply each operation and simplify. When adding and subtracting polynomials, only add and subtract the **coefficients**, or the number part; the variable and exponent stay the same.

ADDING POLYNOMIALS

To add polynomials, you need to add like terms. These terms have the same variable part. For example, the terms $4x^2$ and $3x^2$ both include x^2 terms. To find the sum of like terms, find the sum of the coefficients. Then, keep the same variable part. You can use the distributive property to distribute the plus sign to each term of the polynomial. For example:

$(4x^2 - 5x + 7) + (3x^2 + 2x + 1) =$
$(4x^2 - 5x + 7) + 3x^2 + 2x + 1 =$
$(4x^2 + 3x^2) + (-5x + 2x) + (7 + 1) =$
$7x^2 - 3x + 8$

SUBTRACTING POLYNOMIALS

To subtract polynomials, you need to subtract like terms. To find the difference of like terms, find the difference of the coefficients. Then, keep the same variable part. You can use the distributive property to distribute the minus sign to each term of the polynomial. For example:

$(-2x^2 - x + 5) - (3x^2 - 4x + 1) =$
$(-2x^2 - x + 5) - 3x^2 + 4x - 1 =$
$(-2x^2 - 3x^2) + (-x + 4x) + (5 - 1) =$
$-5x^2 + 3x + 4$

Review Video: Adding and Subtracting Polynomials
Visit mometrix.com/academy and enter code: 124088

MULTIPLYING POLYNOMIALS

In general, multiplying polynomials is done by multiplying each term in one polynomial by each term in the other and adding the results. In the specific case for multiplying binomials, there is a useful acronym, FOIL, that can help you make sure to cover each combination of terms. The **FOIL method** for $(Ax + By)(Cx + Dy)$ would be:

F	Multiply the *first* terms of each binomial	$(\overbrace{Ax}^{first} + By)(\overbrace{Cx}^{first} + Dy)$	ACx^2
O	Multiply the *outer* terms	$(\overbrace{Ax}^{outer} + By)(Cx + \overbrace{Dy}^{outer})$	$ADxy$
I	Multiply the *inner* terms	$(Ax + \overbrace{By}^{inner})(\overbrace{Cx}^{inner} + Dy)$	$BCxy$
L	Multiply the *last* terms of each binomial	$(Ax + \overbrace{By}^{last})(Cx + \overbrace{Dy}^{last})$	BDy^2

Then, add up the result of each and combine like terms: $ACx^2 + (AD + BC)xy + BDy^2$.

For example, using the FOIL method on binomials $(x + 2)$ and $(x - 3)$:

$$\begin{aligned}
&\text{First:} && (\boxed{x} + 2)(\boxed{x} + (-3)) && \rightarrow && (x)(x) &&= x^2 \\
&\text{Outer:} && (\boxed{x} + 2)(x + \boxed{(-3)}) && \rightarrow && (x)(-3) &&= -3x \\
&\text{Inner:} && (x + \boxed{2})(\boxed{x} + (-3)) && \rightarrow && (2)(x) &&= 2x \\
&\text{Last:} && (x + \boxed{2})(x + \boxed{(-3)}) && \rightarrow && (2)(-3) &&= -6
\end{aligned}$$

This results in: $(x^2) + (-3x) + (2x) + (-6)$

Combine like terms: $x^2 + (-3 + 2)x + (-6) = x^2 - x - 6$

Review Video: Multiplying Polynomials
Visit mometrix.com/academy and enter code: 598293

Review Video: Multiplying Terms Using the FOIL Method
Visit mometrix.com/academy and enter code: 854792

Dividing Polynomials

Use long division to divide a polynomial by either a monomial or another polynomial of equal or lesser degree.

When **dividing by a monomial**, divide each term of the polynomial by the monomial.

Review Video: Dividing Monomials
Visit mometrix.com/academy and enter code: 584409

When **dividing by a polynomial**, begin by arranging the terms of each polynomial in order of one variable. You may arrange in ascending or descending order, but be consistent with both polynomials. To get the first term of the quotient, divide the first term of the dividend by the first term of the divisor. Multiply the first term of the quotient by the entire divisor and subtract that product from the dividend. Repeat for the second and successive terms until you either get a remainder of zero or a remainder whose degree is less than the degree of the divisor. If the quotient has a remainder, write the answer as a mixed expression in the form:

$$\text{quotient} + \frac{\text{remainder}}{\text{divisor}}$$

For example, we can evaluate the following expression in the same way as long division:

$$\frac{x^3 - 3x^2 - 2x + 5}{x - 5}$$

$$\begin{array}{r l}
 & \quad x^2 \;\; + 2x \;\; + 8 \\
x - 5 \,\big) & \overline{\; x^3 - 3x^2 \;\; - 2x \;\; + 5} \\
 & \underline{-(x^3 - 5x^2)} \\
 & \qquad\quad 2x^2 - 2x \\
 & \qquad \underline{-(2x^2 - 10x)} \\
 & \qquad\qquad\qquad 8x + 5 \\
 & \qquad\qquad\quad \underline{-(8x - 40)} \\
 & \qquad\qquad\qquad\quad 45
\end{array}$$

$$\frac{x^3 - 3x^2 - 2x + 5}{x - 5} = x^2 + 2x + 8 + \frac{45}{x - 5}$$

Review Video: Dividing Polynomials by Monomials
Visit mometrix.com/academy and enter code: 253551

Review Video: Dividing Trinomials by Binomials
Visit mometrix.com/academy and enter code: 651465

When **factoring** a polynomial, first see whether you can factor out a nontrivial greatest common factor (GCF). For example, the trinomial $3x^5 - 18x^4 + 15x^3$ has a GCF of $3x^3$ since the GCF of 3, 18, and 15 is 3 and the GCF of x^5, x^4, and x^3 is x^3. Factoring out the GCF simplifies the expression to $3x^3(x^2 - 6x + 5)$.

To factor a quadratic trinomial (this comes up frequently), first check whether it is a perfect square trinomial (see bulleted list below). If not, see if you can factor it by trial and error by making clever choices of values for a and b (or a, b, c, and d) in the formulas below (this amounts to trying to use the FOIL mnemonic backwards):

$$x^2 + (a + b)x + ab = (x + a)(x + b)$$
$$(ac)x^2 + (ad + bc)x + bd = (ax + b)(cx + d)$$

For instance, you would try to factor the trinomial $x^2 - 6x + 5$ using the equation $x^2 + (a + b)x + ab = (x + a)(x + b)$. This means that you need to find integers a and b such that $a + b = -6$ and $ab = 5$. Starting with $ab = 5$, you can see that the only ways to write 5 as a product of integers are $(1)(5) = 5$ and $(-1)(-5) = 5$. So, it is easy to see that you want $a = -1$ and $b = -5$ since $a + b = -1 + (-5) = -6$ and $ab = (-1)(-5) = 5$. This tells you that $x^2 - 6x + 5 = (x - 1)(x - 5)$.

For polynomials with four terms (usually a cubic polynomial), sometimes factoring by grouping works: You group the two higher-power terms and the two lower-power terms, factor the GCF out of each group, and then factor out the resulting common binomial factor, if there is one. For example, $x^3 + 5x^2 + 3x + 15 = (x^3 + 5x^2) + (3x + 15) = x^2(x + 5) + 3(x + 5) = (x^2 + 3)(x + 5)$.

Once you have found the factors, write the original polynomial as the product of all the factors. Make sure all of the factors are either monomials, or else linear or irreducible quadratic polynomials (*irreducible* means they have no real zeros, which is easy to check with the quadratic formula). Check your work by multiplying the factors to make sure you get the original polynomial.

> **Review Video: Factoring Out Common Monomial Factors**
> Visit mometrix.com/academy and enter code: 398578
>
> **Review Video: Factoring Trinomials of the Form x^2+bx+c**
> Visit mometrix.com/academy and enter code: 270556

Below are patterns of some special products to remember to help make factoring easier:

- Perfect square trinomials: $x^2 + 2xy + y^2 = (x + y)^2$ or $x^2 - 2xy + y^2 = (x - y)^2$. For example, $x^2 + 10x + 25 = (x + 5)^2$.
- Difference between two squares: $x^2 - y^2 = (x + y)(x - y)$. For example, $x^2 - 9 = (x + 3)(x - 3)$.
- Sum of two cubes: $x^3 + y^3 = (x + y)(x^2 - xy + y^2)$. For example, $x^3 + 27 = (x + 3)(x^2 - 3x + 9)$.
    - Note: the second factor is *not* the same as a perfect square trinomial, so do not try to factor it further.
- Difference between two cubes: $x^3 - y^3 = (x - y)(x^2 + xy + y^2)$. For example, $x^3 - 1000 = (x - 10)(x^2 + 10x + 100)$.
    - Again, the second factor is *not* the same as a perfect square trinomial.
- Perfect cubes: $x^3 + 3x^2y + 3xy^2 + y^3 = (x + y)^3$ and $x^3 - 3x^2y + 3xy^2 - y^3 = (x - y)^3$

> **Review Video: Factoring the Difference of Two Squares**
> Visit mometrix.com/academy and enter code: 128954

Rational and Irrational Expressions

Rational Expressions

Rational expressions are fractions with polynomials in both the numerator and the denominator; the value of the polynomial in the denominator cannot be equal to zero. Be sure to keep track of values that make the denominator of the original expression zero as the final result inherits the same restrictions. For example, a denominator of $x - 3$ indicates that the expression is not defined when $x = 3$ and, as such, regardless of any operations done to the expression, it remains undefined there.

To **add or subtract** rational expressions, first find the common denominator, then rewrite each fraction as an equivalent fraction with the common denominator. Finally, add or subtract the numerators to get the numerator of the answer, and keep the common denominator as the denominator of the answer.

When **multiplying** rational expressions, factor each polynomial and cancel like factors (a factor which appears in both the numerator and the denominator). Then, multiply all remaining factors in the numerator to get the numerator of the product, and multiply the remaining factors in the denominator to get the denominator of the product. Remember: cancel entire factors, not individual terms.

To **divide** rational expressions, take the reciprocal of the divisor (the rational expression you are dividing by) and multiply by the dividend.

Review Video: Rational Expressions
Visit mometrix.com/academy and enter code: 415183

Simplifying Rational Expressions

To simplify a rational expression, factor the numerator and denominator completely. Factors that are the same and appear in the numerator and denominator have a ratio of 1. For example, look at the following expression:

$$\frac{x-1}{1-x^2}$$

The denominator, $(1-x^2)$, is a difference of squares. It can be factored as $(1-x)(1+x)$. The factor $1-x$ and the numerator $x-1$ are opposites and have a ratio of −1. Rewrite the numerator as $-1(1-x)$. So, the rational expression can be simplified as follows:

$$\frac{x-1}{1-x^2}=\frac{-1(1-x)}{(1-x)(1+x)}=\frac{-1}{1+x}$$

Note that since the original expression is only defined for $x \neq \{-1, 1\}$, the simplified expression has the same restrictions.

Review Video: Reducing Rational Expressions
Visit mometrix.com/academy and enter code: 788868

Review Video: Simplifying Algebraic Expressions with Parentheses
Visit mometrix.com/academy and enter code: 850843

Irrational Expressions

Irrational expressions are mathematical expressions that contain an irrational number and cannot be simplified into a rational form. Usually, this includes expressions that contain radicals or constants such as π. Most commonly, you will encounter these in the forms of expressions containing roots of non-perfect squares.

Basic Operations on Radical Expressions

To add or subtract radical numbers, the numbers within the radicals must match, similar to finding a common denominator in fractions:

$$a\sqrt{x}+b\sqrt{x}=(a+b)\sqrt{x}$$

$$a\sqrt{x}-b\sqrt{x}=(a-b)\sqrt{x}$$

To multiply radicals, the numbers outside the radical are multiplied together and the numbers inside the radical are multiplied together:

$$a\sqrt{x}\times b\sqrt{y}=ab\sqrt{xy}$$

To divide radicals, the radical must be eliminated from the denominator by multiplying both numerator and denominator by a value that will make the denominator a rational number:

$$\frac{a\sqrt{x}}{b\sqrt{y}} = \frac{a\sqrt{x}\sqrt{y}}{b\sqrt{y}\sqrt{y}} = \frac{a\sqrt{xy}}{by}$$

Review Video: Adding and Subtracting Radical Expressions
Visit mometrix.com/academy and enter code: 752176

EXAMPLE

To solve $\frac{(3\sqrt{6})(2\sqrt{3})}{4\sqrt{5}}$, we first multiply the numerator, inside the radical and out: $3 \times 2\sqrt{6 \times 3} = 6\sqrt{18} = 18\sqrt{2}$. To divide, we multiply both numerator and denominator by a value that will eliminate the radical:

$$\frac{18\sqrt{2} \times \sqrt{5}}{4\sqrt{5} \times \sqrt{5}} = \frac{18\sqrt{10}}{4 \times 5} = \frac{18\sqrt{10}}{20} = \frac{9\sqrt{10}}{10}$$

ALGEBRAIC THEOREMS

According to the **fundamental theorem of algebra**, every non-constant, single-variable polynomial has exactly as many roots as the polynomial's highest exponent. For example, if x^4 is the largest exponent of a term, the polynomial will have exactly 4 roots. However, some of these roots may have multiplicity or be complex numbers. For instance, in the polynomial function $f(x) = x^4 - 4x + 3$, the only real root is 1, though it has multiplicity of 2 – that is, it occurs twice. The other two roots, $(-1 - i\sqrt{2})$ and $(-1 + i\sqrt{2})$, are complex, consisting of both real and non-real components.

The **remainder theorem** is useful for determining the remainder when a polynomial is divided by a binomial. The remainder theorem states that if a polynomial function $f(x)$ is divided by a binomial $x - a$, where a is a real number, the remainder of the division will be the value of $f(a)$. If $f(a) = 0$, then a is a root of the polynomial.

The **factor theorem** is related to the remainder theorem and states that if $f(a) = 0$ then $(x - a)$ is a factor of the function.

According to the **rational root theorem,** any rational root of a polynomial function $f(x) = a_nx^n + a_{n-1}x^{n-1} + \cdots + a_1x + a_0$ with integer coefficients will, when reduced to its lowest terms, be a positive or negative fraction such that the numerator is a factor of a_0 and the denominator is a factor of a_n. For instance, if the polynomial function $f(x) = x^3 + 3x^2 - 4$ has any rational roots, the numerators of those roots can only be factors of 4 (1, 2, 4), and the denominators can only be factors of 1 (1). The function in this example has roots of 1 (or $\frac{1}{1}$) and -2 (or $\frac{-2}{1}$).

QUADRATICS

SOLVING QUADRATIC EQUATIONS

A quadratic equation is an equation that can be written (possibly after simplification) in the form $ax^2 + bx + c = 0$. Thus, the **solutions** of this equation are precisely the **zeros** of the quadratic polynomial $P(x) = ax^2 + bx + c$. On the graph of this polynomial the zeros, if any, appear as

x-intercepts. There are several ways to find these solutions including the quadratic formula, factoring, completing the square, and graphing the function.

Review Video: Quadratic Equations Overview
Visit mometrix.com/academy and enter code: 476276

Review Video: Solutions of a Quadratic Equation on a Graph
Visit mometrix.com/academy and enter code: 328231

Quadratic Formula

The **quadratic formula** gives the zeros of a quadratic polynomial. It always works, but it is sometimes a little harder to use than other methods. To use it to solve a quadratic equation, rewrite the equation in the form $ax^2 + bx + c = 0$, where a, b, and c are coefficients. Now, as explained above, the solutions of the equation are the zeros of the quadratic polynomial $P(x) = ax^2 + bx + c$. To find them, substitute the values of a, b, and c into the Quadratic Formula:

$$x = \frac{-b \pm \sqrt{b^2 - 4ac}}{2a}$$

After simplification this formula produces two, one, or zero real solutions, depending on whether the **discriminant** (the number $b^2 - 4ac$ under the radical) is positive, zero, or negative. It is a good practice to check each solution by substituting it into the original equation. Incidentally, if the discriminant is negative, then the equation does have two complex solutions, but you often ignore these as meaningless in real-world settings.

Review Video: Using the Quadratic Formula
Visit mometrix.com/academy and enter code: 163102

Factoring

To solve a quadratic equation by factoring, begin by rewriting the equation in the standard form, $ax^2 + bx + c = 0$. In the important special case that $a = 1$, the goal of factoring is to find numbers f and g such that $x^2 + bx + c = (x + f)(x + g) = x^2 + (f + g)x + fg$. In other words, you want to choose f and g to make $fg = c$ and $f + g = b$. To do this, find pairs of numbers (factors) whose product is c and look for a pair whose sum is b.

For example, suppose you want to find the solutions of the equation $x^2 + 6x - 16 = 0$ by factoring. Here $b = 6$ and $c = -16$. First, you find the pairs of numbers whose product is -16. These are -4 and 4, -8 and 2, -2 and 8, -1 and 16, and 1 and -16. The pair -2 and 8 has a sum of 6. This means $f = -2$ and $g = 8$. So, the factorization is $x^2 + 6x - 16 = (x + f)(x + g) = (x - 2)(x + 8)$, allowing you to rewrite the original equation as $(x - 2)(x + 8) = 0$. The only way for a product to equal zero is for one of the factors to equal zero; so, either $x - 2 = 0$ (in which case $x = 2$) or $x + 8 = 0$ (in which case $x = -8$). Thus, the equation has the solution $x = 2$ or $x = -8$.

In the case that $a \neq 1$, you can attempt to factor the quadratic polynomial $ax^2 + bx + c$ in the form $(mx + f)(nx + g)$ by a similar trial-and-error procedure, but the work tends to be much harder.

Review Video: Factoring Quadratic Equations
Visit mometrix.com/academy and enter code: 336566

COMPLETING THE SQUARE

The technique of completing the square comes from a simple observation: Suppose you have the expression $x^2 + bx$. If you take half the linear coefficient, b, square it, and add the result to the expression, the result is always a perfect square trinomial:

$$x^2 + bx + \left(\frac{b}{2}\right)^2 = \left(x + \frac{b}{2}\right)^2$$

For example, if you begin with $x^2 + 6x$ and add the square of half of 6 (half of 6 is 3, and $3^2 = 9$), then you get a perfect square trinomial:

$$x^2 + 6x + 9 = (x + 3)^2$$

This also works if b is negative. For instance, if you begin with the expression $x^2 - 10x$ and complete the square by adding 25 (half of -10 is -5, and $(-5)^2 = 25$), then you get another perfect square trinomial:

$$x^2 - 10x + 25 = (x - 5)^2$$

Now, suppose you want to solve the equation $x^2 + bx + c = 0$. Subtract c from both sides, to get the equation $x^2 + bx = -c$. Complete the square on the left side by adding $(b/2)^2$, but also add this same term to the right side so that the new equation is equivalent to the old one:

$$x^2 + bx + \left(\frac{b}{2}\right)^2 = -c + \left(\frac{b}{2}\right)^2$$
$$\left(x + \frac{b}{2}\right)^2 = -c + \left(\frac{b}{2}\right)^2$$

Take square roots of both sides:

$$x + \frac{b}{2} = \pm\sqrt{-c + \left(\frac{b}{2}\right)^2}$$

Remember to include $\pm$ since every positive number has both a positive and a negative square root. Subtract $b/2$ from both sides to isolate the variable x to finish the problem:

$$x = -\frac{b}{2} \pm \sqrt{-c + \left(\frac{b}{2}\right)^2}$$

This may sound complicated, but in practice it is not hard. For example, suppose you want to solve the equation $x^2 + 6x - 16 = 0$ by completing the square. First, add 16 to both sides:

$$x^2 + 6x = 16$$

Half of 6 is 3 and $3^2 = 9$, so complete the square by adding 9 to both sides of the equation:

$$x^2 + 6x + 9 = 16 + 9$$
$$(x + 3)^2 = 25$$

Now take square root of both sides, remembering to include the $\pm$:

$$\sqrt{(x+3)^2} = \pm\sqrt{25}$$
$$x + 3 = \pm 5$$
$$x = -3 \pm 5$$

So we see that the two solutions to the equation are $x = 2$ and $x = -8$.

> **Review Video: Completing the Square**
> Visit mometrix.com/academy and enter code: 982479

Using Given Solutions to Find a Quadratic Equation

To find a quadratic equation with given numbers as solutions, simply find a quadratic polynomial that has those numbers as zeros and set that polynomial equal to zero. This is easy because a polynomial has the number p as a zero precisely when it has the binomial $x - p$ as a factor. Thus, for instance, to find a quadratic polynomial with zeros at $x = 3$ and $x = -5$, construct the polynomial $P(x) = (x-3)(x-(-5)) = (x-3)(x+5) = x^2 + 2x - 15$. Setting this equal to zero produces an equation, $x^2 + 2x - 15 = 0$, whose solutions are $x = 3$ and $x = -5$.

Of course, any constant multiple $P(x) = a(x-3)(x+5)$ will also have the same zeros. For instance, if you choose $a = 4$, then you get the polynomial $P(x) = 4(x-3)(x+5) = 4x^2 + 8x - 60$, which also has zeros at $x = 3$ and $x = -5$. From this you can get another equation, $4x^2 + 8x - 60 = 0$, with solutions $x = 3$ and $x = -5$.

Functions

Basics of Functions

Definition of a Function

A function is a rule that assigns to every number in a given set (called the **domain**) exactly one corresponding value. For example, if our domain is the set $\{-2,1,2,3\}$, we can define a function by assigning to each number its square. This function assigns to -2 the value 4, to 1 the value 1, to 2 the value 4, and to 3 the value 9 (since $(-2)^2 = 4$, $1^2 = 1$, $2^2 = 4$, and $3^2 = 9$). The set of all the values assigned by a function is the **range** of the function. The range of the function in our example is the set $\{1, 4, 9\}$. We may think of a function as a kind of machine: we give it a number as an input, and it uses its rule to produce a number as an output. In the squaring function above, the input 3 produces the output 9.

> **Review Video: What is a Function?**
> Visit mometrix.com/academy and enter code: 784611

Function Notation

We usually name a function by a letter, often the letter f (for *function*—if we need to talk about more than one function, we name the second one g, the third one h, etc.). To specify the value (the output) corresponding to a particular number in the domain (the input), we write the function letter followed by the input number in parentheses. For instance, in the example above the notation $f(3)$ means the value that the function assigns to the number 3, namely 9—that is, $f(3) = 9$. We read the symbols $f(3)$ as, "f of 3," and we call 3 the **argument** of the function and 9 the **value** of the function (so *argument* means *input* and *value* means *output*).

Using function notation we can define the squaring function above by listing the values the function assigns to each argument in the domain: $f(-2) = 4$. $f(1) = 1$, $f(2) = 4$, and $f(3) = 9$. More efficiently, we can define the function by the single equation $f(x) = x^2$, which says that if x is a number from the domain, then we calculate the value assigned to it by substituting the number x in the formula x^2. For instance, we calculate $f(5) = 5^2 = 25$. Similarly, if we define a function g by the equation $g(x) = x^2 - 4x + 7$, then we calculate the value $g(3)$ by substituting 3 for each x in the formula: $g(3) = 3^2 - 4 \cdot 3 + 7 = 9 - 12 + 7 = 4$.

Other Ways to Define Functions

Instead of denoting the value of the function by $f(x)$, sometimes we simply use another letter, usually y. For instance, instead of defining the squaring function by the equation $f(x) = x^2$, we might use the equation $y = x^2$. In this case, we refer to x (the input) as the **independent variable** and y (the output) as the **dependent variable** because the value, y, depends on the number we choose for x.

A formula (with y or $f(x)$) is the most common way to define a function; but sometimes, if the domain is small enough, we prefer to list explicitly the possible inputs and their corresponding outputs. Some ways of doing this appear above, but a more common approach is to put the input-output pairs in a table. For instance, we can define the squaring function above by the table

x	−2	1	2	3
y	4	1	4	9

We see that the domain of this function is the set of all numbers in the x-row and the range is the set of all numbers in the y-row. We note that numbers cannot repeat in the x-row (because a function assigns exactly one value to each argument in the domain) but they can repeat in the y-row (because the function can assign the same value to multiple arguments—for instance, the number 4 appears twice in the y-row).

We can also define a function by writing the inputs and corresponding outputs as ordered pairs of x- and y-values. For instance, we can write the squaring function above as the set of ordered pairs $\{(-2,4), (1,1), (2,4), (3,9)\}$. Further, by treating these ordered pairs as coordinates and plotting the corresponding points on the coordinate plane, we get the **graph** of the function:

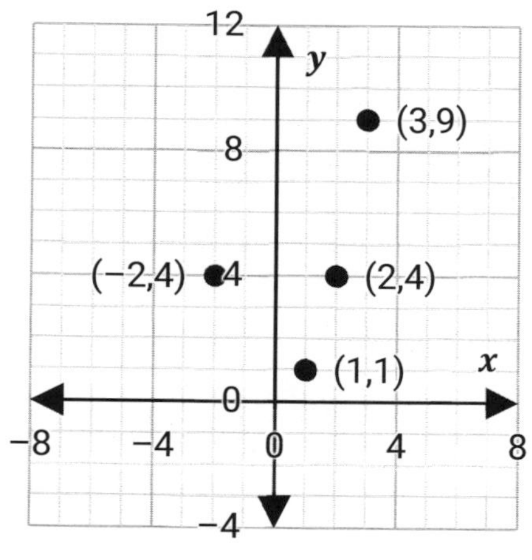

Turning this around, we can potentially use a graph to define a function, namely the function consisting of the coordinate pairs of all the points in the graph. This always works unless the graph

has two points with the same x-coordinate (because then the function would assign two different y-values to the same x). It is easy to detect such points: because they have the same x-coordinate, a vertical line passes through both of them. Thus, a graph always defines a function unless it is possible to draw a vertical line that intersects the graph in two or more points. We call this condition the **vertical line test**. For example, if our graph is a circle, then by the Vertical Line Test the graph does not define a function because there are vertical lines that will intersect the circle in two different points.

MORE ON DOMAINS AND RANGES

When we define a function by a formula and do not specify the domain, then by default the domain consists of all real numbers for which the formula produces an answer. For instance, suppose we define a function f by the formula $f(x) = 1/x$. If $x = 0$, then $1/x = 1/0$, which is undefined. But if x is any other real number, then we can calculate the value of $1/x$. So, the default domain of this function is all real numbers except zero. Because of this domain convention, the graph of a function defined by a formula usually consists of infinitely many points that "connect to" each other in a way that produces a line or curve (see examples below) rather than the isolated points we see in the squaring function above.

If we have the graph of a function, its domain consists of all numbers on the x-axis with corresponding points on the graph and its range consists of all numbers on the y-axis with corresponding points on the graph. For example, consider the function $f(x) = x^2 + 3$:

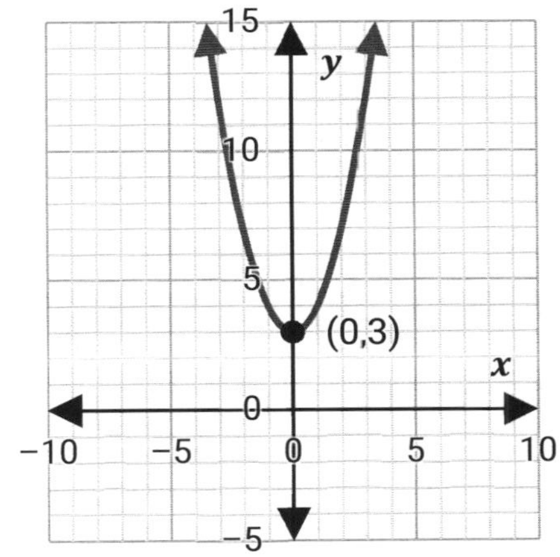

Since the graph continues infinitely to the left and right beyond what we can see, every point on the x-axis has a corresponding point on the graph; so, the domain of this function is all real numbers. On the other hand, the lowest point on this graph has a y-value of 3, and the graph passes through all higher y-values. So, the range of this function is all real numbers greater than or equal to 3, which we can denote algebraically by $y \geq 3$ or, using interval notation, by $[3, \infty)$.

Review Video: <u>How to Find Domain and Range</u>
Visit mometrix.com/academy and enter code: 778133

Review Video: <u>Domain and Range of Quadratic Functions</u>
Visit mometrix.com/academy and enter code: 331768

MONOTONIC AND EVEN/ODD FUNCTIONS

A function, f, is **increasing** if it always assigns larger values to larger arguments. It is **decreasing** if it always assigns smaller values to larger arguments. That is, f is increasing if $a < b$ always guarantees $f(a) < f(b)$, and it is decreasing if $a < b$ always guarantees $f(a) > f(b)$. The graph of an increasing function consistently rises from left to right, and the graph of a decreasing function consistently falls from left to right. For example, the function $f(x) = 2x$ is an increasing function because doubling a larger number always gives us a larger result than doubling a smaller number. The graph of $f(x) = 2x$ is a line with slope $m = 2$, which, as we expect, rises from left to right. We call a function **monotonic** if it is either increasing or decreasing.

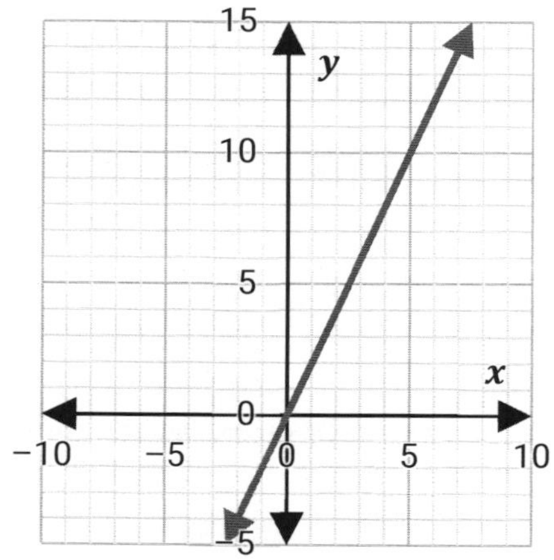

A function, f, is **even** if changing the sign of its argument produces the same value. It is **odd** if changing the sign of its argument produces the same value except with the opposite sign. That is, f is even if $f(-x) = f(x)$ and odd if $f(-x) = -f(x)$ for every argument x. The function $f(x) = x^2 + 3$ is even because substituting opposite arguments always produces the same value. For instance, $f(5) = 28$ and $f(-5) = 28$ because $5^2 + 3 = 25 + 3 = 28$ and $(-5)^2 + 3 = 25 + 3 = 28$. The function $f(x) = 2x$ is odd because substituting opposite arguments always produces opposite values. For instance, $f(10) = 20$ and $f(-10) = -20$ because $2(10) = 20$ and $2(-10) = -20$. The graph of an even function is always symmetric with respect to the y-axis, making the left and right halves of the graph mirror images of each other, as in the graph of the even function $f(x) = x^2 + 3$ above. The graph of an odd function is always symmetric with respect to the origin. This means that if we rotate the graph 180° around the origin (think of sticking a pin through the origin on a sheet of graph paper and rotating the paper halfway around) the graph looks the same, as in the graph of the odd function $f(x) = 2x$ above.

It is worth noting that most functions are neither increasing nor decreasing (that is, they are not monotonic) and most functions are neither even nor odd. For example, the function $f(x) = x^2 -$

x is neither increasing nor decreasing and neither even nor odd: its graph neither rises nor falls consistently, and it is symmetric with respect to neither the y-axis nor the origin.

Review Video: Even and Odd Functions
Visit mometrix.com/academy and enter code: 278985

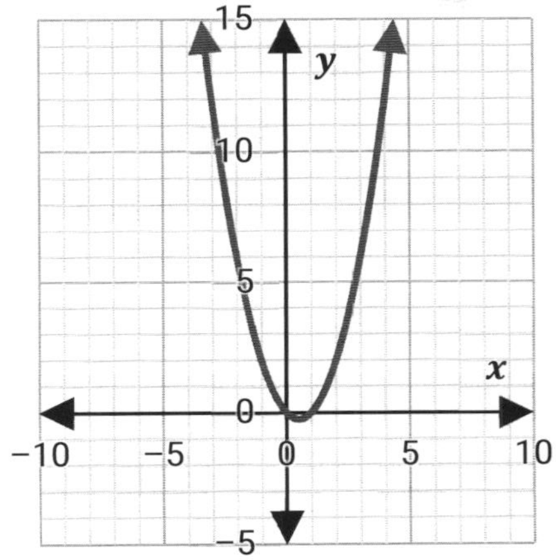

Invertible (One-to-One) Functions

A function, f, is one-to-one if it never assigns the same value to different arguments—that is, if $f(a)$ and $f(b)$ are different whenever a and b are different. The graph of a one-to-one function never has two points that lie on the same horizontal line because such points would have different x-values but the same y-value. Thus, a function is one-to-one if it is impossible to draw a horizontal line that intersects its graph in more than one point. We call this condition the **horizonal line test**. For example, the graph of the function $f(x) = 2x$ above is a line that rises from left to right. Every horizontal line intersects this line in exactly one point, so the function $f(x) = 2x$ is one-to-one. This is also clear without the graph because it is impossible to double two different numbers and get the same answer.

When a function, f, is one-to-one, it is possible to define its inverse function, f^{-1}, that "undoes" what f does, assigning to each output from f the input that produced it. That is, for each x in the domain of f, if $y = f(x)$, then $f^{-1}(y) = x$. For example, the inverse of the function $f(x) = 2x$ above is $f^{-1}(y) = y/2$. So, for instance, $f(5) = 2 \cdot 5 = 10$, and $f^{-1}(10) = 10/2 = 5$ (and similarly for every other value of x). Thus, the domain of f^{-1} is the range of f and vice versa. If a function, f, has an inverse, we say that f is **invertible**. Since a function has an inverse precisely when it is one-to-one, the terms *invertible* and *one-to-one* are synonyms.

If f is an invertible function defined by a formula, then to find its inverse we simply write the equation $y = f(x)$ and solve it for x (that is, we isolate the x). The result will be the equation $f^{-1}(y) = x$. For instance, starting with the function $f(x) = 2x$, we write $y = 2x$ and isolate the x by dividing both sides of the equation by 2. This gives us $y/2 = x$, so we know that $f^{-1}(y) = y/2$. Although this procedure is theoretically simple, in practice the algebra can be difficult.

Common Functions

Certain functions and certain kinds of functions are particularly useful, coming up frequently in mathematics and its applications. Once we know some basic function terminology and concepts, it is useful to begin developing a mental library of the most common and useful functions.

Review Video: Common Functions
Visit mometrix.com/academy and enter code: 629798

Constant Functions

A function of the form $f(x) = a$, where a is a real number, is a **constant function**. This function assigns the same value, a, to every real argument x. For instance, given the constant function $f(x) = 5$, we have $f(2) = 5$, $f(100) = 5$, and $f(-7.1) = 5$.

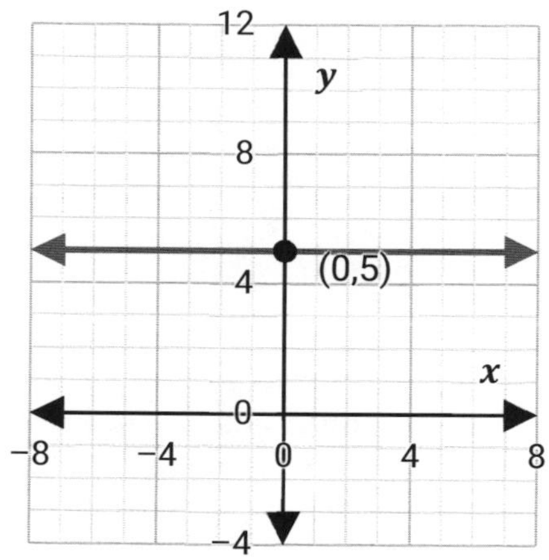

The domain of a constant function is the set of all real numbers, and the range is the set containing the single number a. Its graph is a horizontal line passing through the number $y = a$ on the y-axis (we call the number at which a function's graph intersects the y-axis the **y-intercept** of the function).

The Identity Function

The function $f(x) = x$ is the **identity function**. Its value always equals its argument. Thus, for instance, $f(2) = 2$, $f(100) = 100$, and $f(-7.1) = -7.1$.

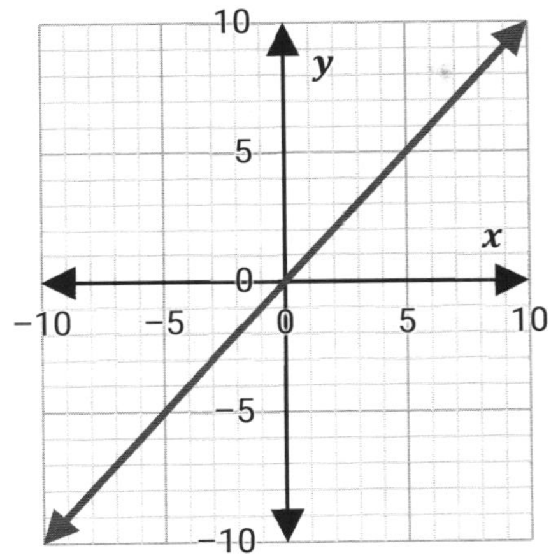

Its domain and range are the set of all real numbers. It is both an increasing function and an odd function. Its graph is a line that passes through the origin and rises from left to right at a 45° angle to the horizontal. Since it passes through the origin, its y intercept is $y = 0$ and it also has an **x-intercept** (a number at which the function's graph intersects the x-axis) of $x = 0$.

Linear Functions

A function of the form $f(x) = ax + b$, where a and b are real numbers (with $a \neq 0$), is a **linear function** (the identity function is a linear function with $a = 1$ and $b = 0$). Its domain and range are the set of all real numbers. Its graph is a line (the word *linear* contains the root word *line*) with one x-intercept (at $x = -b/a$), with a y-intercept at $y = b$, and with a direction and steepness that depend on the coefficient a, which we call the **slope**. Specifically, the slope a is the amount the y-value increases for each increase of 1 in the x-value. Thus, for $a > 0$, the line rises from left to right (making f an increasing function), and larger values of a produce steeper ascents. Similarly, for $a < 0$, the line falls from left to right (making f a decreasing function), and smaller (more negative) values of a produce steeper descents. For instance, the graph of the linear function

$f(x) = (1/2)x + 3$ is a line that passes through the point $y = 3$ on the y-axis and that rises by $1/2$ unit for every unit that x increases.

Review Video: Linear Functions
Visit mometrix.com/academy and enter code: 200735

Review Video: Graphing Linear Functions
Visit mometrix.com/academy and enter code: 699478

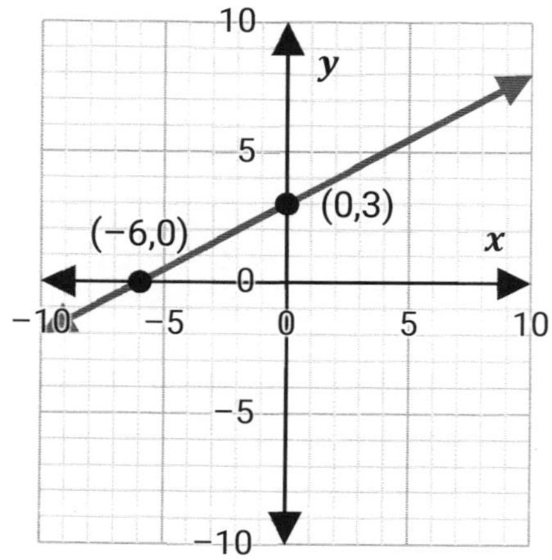

In many contexts it is standard to use the letter m for slope and thus to write the general form of a linear function as $f(x) = mx + b$, known as **slope-intercept form**.

THE SQUARING FUNCTION

The function $f(x) = x^2$ is the **squaring function**.

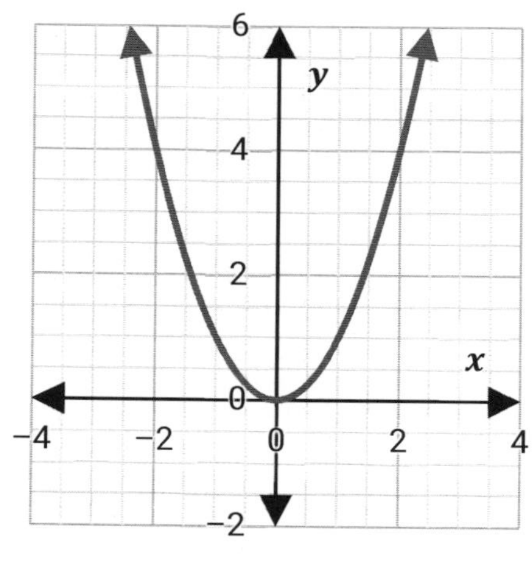

Its graph is U-shaped, opening upward as shown, a shape known as a **parabola**. It has a lowest point, its **vertex**, at the origin, which is also the location of its single x-intercept and single y-intercept. Thus, its **minimum** is $y = 0$, its domain is the set of all real numbers, and its range is the set of nonnegative real numbers (that is, $y \geq 0$). It is an even function and thus symmetric with respect to the y-axis (which we call the **axis of symmetry**), meaning that the left half of the graph is the mirror image of the right half, with the mirror standing on the y-axis.

Quadratic Functions

A function of the form $f(x) = ax^2 + bx + c$, where a, b, and c are real numbers (with $a \neq 0$), is a **quadratic function** (the squaring function is a quadratic function with $a = 1, b = 0$, and $c = 0$). Its domain is the set of all real numbers, and its graph is a parabola. It is symmetric with respect to its axis of symmetry, the vertical line $x = -b/(2a)$. If $a > 0$, the parabola opens upward, so that its vertex is at its lowest point (its minimum) and its range consists of all real numbers greater than or equal to this minimum y-value. If $a < 0$, the parabola opens downward, so that its vertex is at its highest point (its maximum) and its range consists of all real numbers less than or equal to this maximum y-value. Its y-intercept is $y = c$ since $f(0) = c$, and it may have zero, one, or two x-intercepts. For example, the function $f(x) = x^2 - 6x + 5$ has $a = 1, b = -6$, and $c = 5$.

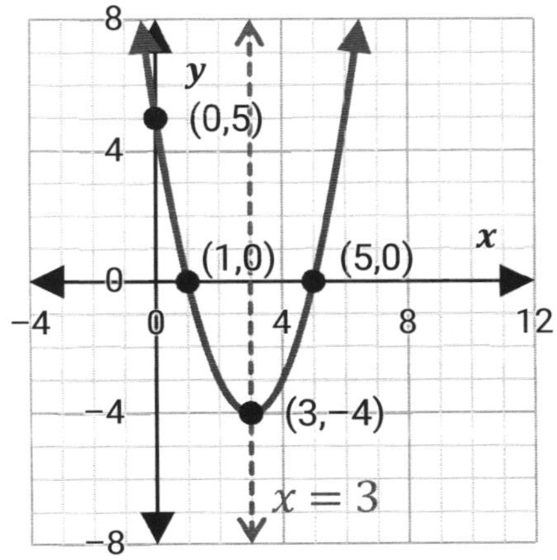

Its graph opens upward (because $a > 0$) and its axis of symmetry is the vertical line $x = 3$ (since $-b/(2a) = -(-6)/(2 \cdot 1) = 3$). Its y-intercept is at $y = 5$. It turns out to have its vertex at the point $(3, -4)$, making its minimum value $y = -4$. So, its domain is the set of all real numbers, and its range is $y \geq -4$. It also turns out to have two x-intercepts, at $x = 1$ and at $x = 5$ (since $f(1) = 0$ and $f(5) = 0$).

Polynomial Functions

A function of the form $f(x) = a^n x^n + a^{n-1} x^{n-1} + \cdots + a_2 x^2 + a_1 x + a_0$, where n is a whole number and $a_0, a_1, a_2, \ldots a_{n-1}, a_n$ are real numbers, is a **polynomial function of degree *n***. Its domain is the set of all real numbers (it is complicated to describe its range in general), and its y-intercept is $y = a_0$ (since $f(0) = a_0$). Constant functions, linear functions, and quadratic functions are polynomial functions of degrees 0, 1, and 2, respectively. In general, a polynomial function of degree n has up to n zeros (x-intercepts) and up to $n - 1$ "bends." For example, the fourth degree polynomial function $f(x) = x^4 - 11x^3 + 41x^2 - 61x + 30$, whose graph appears here, has four x-intercepts (at $x = 1, x = 2, x = 3$, and $x = 5$) and three "bends," and its y-intercept (not visible on the graph) is at $y = 30$.

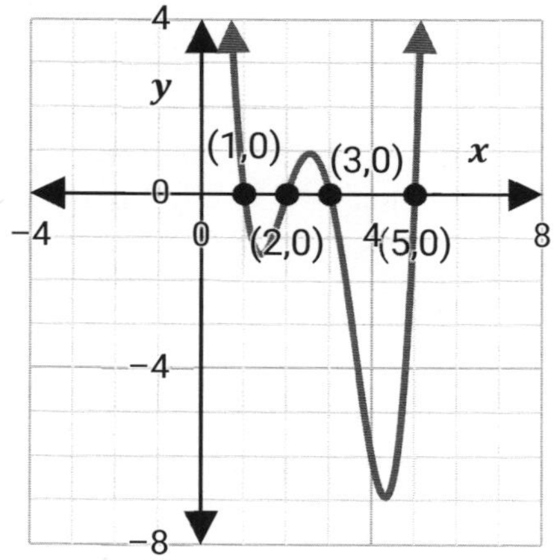

Rational Functions

A function of the form $f(x) = P(x)/Q(x)$, where P and Q are polynomials, is a rational function (we note that the word *rational* includes the root word *ratio*, indicating that a rational function is a ratio of polynomial functions). The domain of a rational function is all real numbers except the zeros of $Q(x)$ since division by zero is undefined (the range can be difficult to describe in general). Its y-intercept is $f(0)$, if this is defined; and its x-intercepts are the zeros of $P(x)$ that are in the domain of f, if there are any. A rational function may also have vertical asymptotes (vertical lines that the graph approaches without crossing) and a horizontal asymptote (a horizontal line that the curve approaches as x becomes very small or very large (toward the left and right edges of the graph). For example, the rational function $f(x) = (2x^2 + x - 1)/(x^2 + x - 2)$ has as its domain the set of all real numbers except $x = -2$ and $x = 1$ (since those numbers make the denominator zero).

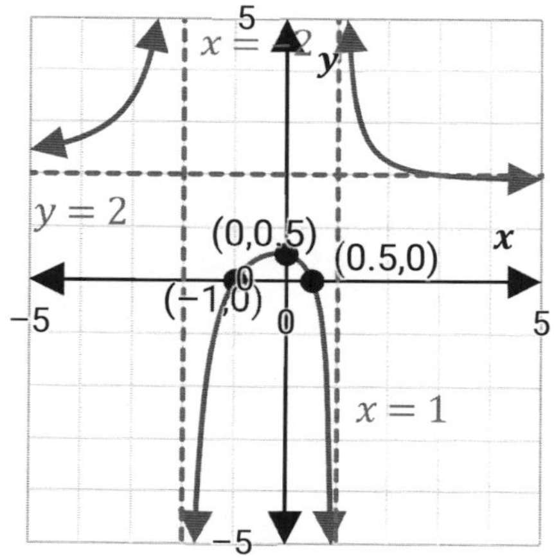

It has a y-intercept of $y = 1/2$ since $f(0) = (-1)/(-2) = 1/2$, and it has x-intercepts at $x = -1$ and at $x = 1/2$ since those numbers make the numerator zero. It has vertical asymptotes at $x = -2$ and $x = 1$ (not coincidentally, these are the numbers omitted from the domain) and a horizontal asymptote at $y = 2$. It is important to note that vertical asymptotes cannot be crossed in rational functions, but horizontal asymptotes can be crossed if the function tends near the asymptote at infinity and does not go past all possible turning points.

> **Review Video: Simplifying Rational Polynomial Functions**
> Visit mometrix.com/academy and enter code: 351038
>
> **Review Video: Horizontal Asymptotes**
> Visit mometrix.com/academy and enter code: 747796

THE SQUARE ROOT FUNCTION

The function $f(x) = \sqrt{x}$ is the square root function.

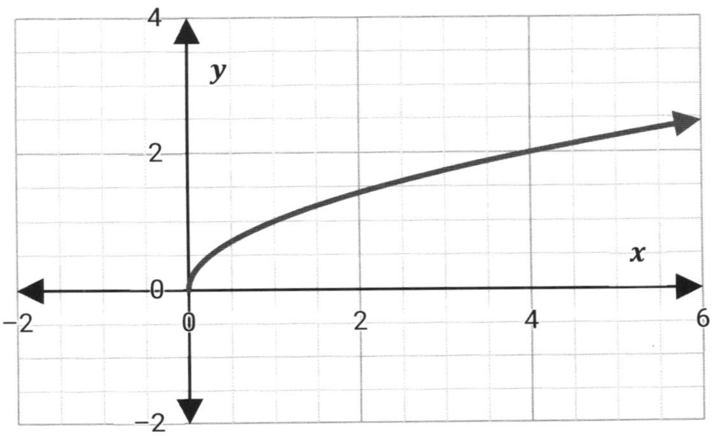

It is an increasing function, and its domain and range are both the set of all nonnegative real numbers. It has one x-intercept and one y-intercept, both appearing at the origin. Its graph is the upper half of a parabola opening to the right. The square root function is the inverse of the squaring function with domain restricted to the nonnegative real numbers (that is, $f(x) = x^2$ for $x \geq 0$).

PIECEWISE-DEFINED FUNCTIONS

As the name suggests, a **piecewise-defined function** (or, simply, a **piecewise function**) is a function defined by different rules on different pieces of the domain. We define such a function using the following form:

Function Name	Rule to Apply	Piece of the Domain on Which the Rule Applies
$f(x) =$	Rule 1,	First Piece of the Domain
	Rule 2,	Second Piece of the Domain
	Rule 3,	Third Piece of the Domain
	etc.,	etc.

The pieces of the domain should not overlap, and together they should cover the whole domain. For example, we might craft a piecewise-defined function by

$$f(x) = \begin{cases} x^2, & \text{if } x < 2 \\ 3x - 5, & \text{if } x \geq 2 \end{cases}$$

The two pieces of the domain—namely, $x < 2$ and $x \geq 2$—do not overlap, and together they include all real numbers. To evaluate the function for a particular argument x, we determine which piece of the domain includes x and then apply the corresponding rule. For instance, to find $f(4)$, we note that $4 \geq 2$; so, we apply the rule $3x - 5$ to get the value $f(4) = 3 \cdot 4 - 5 = 7$. Similarly, to find $f(-6)$, we note that $-6 < 2$; so, we apply the rule x^2 to get the value $f(-6) = (-6)^2 = 36$.

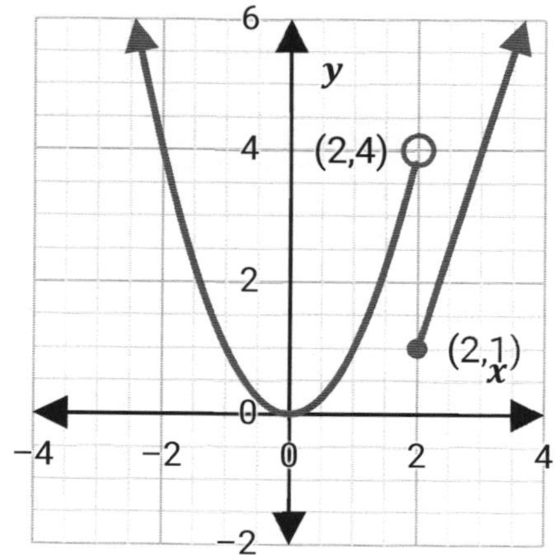

To graph this function, we sketch the graph of the parabola $y = x^2$ on the part of the plane where $x < 2$ and we sketch the line $y = 3x - 5$ on the part of the plane where $x \geq 2$. This produces a graph with a jump at $x = 2$ (a discontinuity—piecewise-defined functions are useful for producing graphs with discontinuities). We plot an open circle at the point (2,4), the end of the left part of the graph, to show that this point is not part of the graph. And we plot a solid dot at the point (2,1), the start of the right part of the graph, to show that this point *is* part of the graph.

Review Video: Piecewise Functions
Visit mometrix.com/academy and enter code: 707921

The Absolute Value Function

A particularly useful piecewise-defined function is the absolute value function. It is so important that instead of naming it $f(x)$ or $g(x)$, we denote it using the special notation $|x|$. Its definition is

$$|x| = \begin{cases} -x, & \text{if } x < 0 \\ x, & \text{if } x \geq 0 \end{cases}$$

For instance, $|8| = 8$ (since $8 \geq 0$) and $|-5| = -(-5) = 5$, since $-5 < 0$. So, the absolute value function acts like the identity function for nonnegative numbers (it leaves them unchanged), and it gives the opposite of negative numbers (it effectively strips off the minus sign). Thus, we can think of the absolute value of a real number as its distance from zero on the number line, without taking into consideration whether the number is larger than or smaller than zero. For instance, $|-3| = 3$ and $|3| = 3$, showing that both -3 and 3 are three units away from zero. The absolute value function is an even function with a V-shaped graph that looks like the line $y = x$ (the identity function) on the right "half" of the plane (for $x \geq 0$) and the line $y = -x$ on the left "half" of the plane (for $x < 0$).

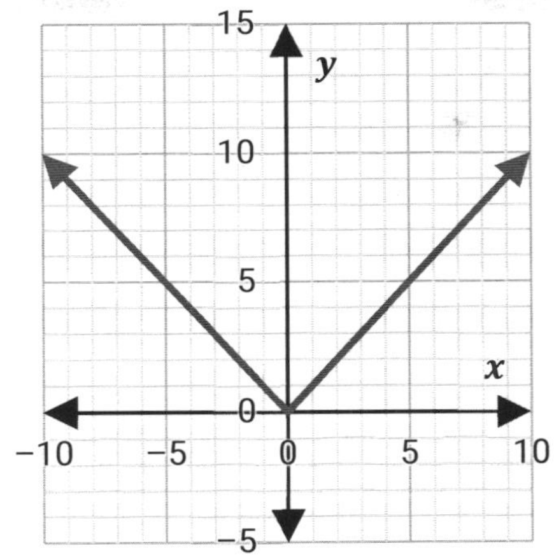

Working with Functions

Manipulation of Functions

Translation occurs when values are added to or subtracted from the x- or y-values. If a constant is added to the y-portion of each point, the graph shifts up. If a constant is subtracted from the y-portion of each point, the graph shifts down. This is represented by the expression $f(x) \pm k$, where k is a constant. If a constant is added to the x-portion of each point, the graph shifts left. If a constant is subtracted from the x-portion of each point, the graph shifts right. This is represented by the expression $f(x \pm k)$, where k is a constant.

Stretching, compression, and reflection occur when different parts of a function are multiplied by different groups of constants. If the function as a whole is multiplied by a real number constant greater than 1, $(k \times f(x))$, the graph is stretched vertically. If k in the previous equation is greater than zero but less than 1, the graph is compressed vertically. If k is less than zero, the graph is reflected about the x-axis, in addition to being either stretched or compressed vertically if k is less than or greater than –1, respectively. If instead, just the x-term is multiplied by a constant greater than 1 $(f(k \times x))$, the graph is compressed horizontally. If k in the previous equation is greater than zero but less than 1, the graph is stretched horizontally. If k is less than zero, the graph is reflected about the y-axis, in addition to being either stretched or compressed horizontally if k is greater than or less than –1, respectively.

Review Video: Manipulation of Functions
Visit mometrix.com/academy and enter code: 669117

Applying the Basic Operations to Functions

For each of the basic operations, we will use these functions as examples: $f(x) = x^2$ and $g(x) = x$.

To find the sum of two functions f and g, assuming the domains are compatible, simply add the two functions together: $(f + g)(x) = f(x) + g(x) = x^2 + x$.

To find the difference of two functions f and g, assuming the domains are compatible, simply subtract the second function from the first: $(f - g)(x) = f(x) - g(x) = x^2 - x$.

To find the product of two functions f and g, assuming the domains are compatible, multiply the two functions together: $(f \times g)(x) = f(x) \times g(x) = x^2 \times x = x^3$.

To find the quotient of two functions f and g, assuming the domains are compatible, divide the first function by the second: $\left(\frac{f}{g}\right)(x) = \frac{f(x)}{g(x)} = \frac{x^2}{x} = x\,; x \neq 0$.

The example given in each case is fairly simple, but on a given problem, if you are looking only for the value of the sum, difference, product, or quotient of two functions at a particular x-value, it may be simpler to solve the functions individually and then perform the given operation using those values.

The composite of two functions f and g, written as $(f \circ g)(x)$ simply means that the output of the second function is used as the input of the first. This can also be written as $f(g(x))$. In general, this can be solved by substituting $g(x)$ for all instances of x in $f(x)$ and simplifying. Using the example functions $f(x) = x^2 - x + 2$ and $g(x) = x + 1$, we can find that $(f \circ g)(x)$ or $f(g(x))$ is equal to $f(x + 1) = (x + 1)^2 - (x + 1) + 2$, which simplifies to $x^2 + x + 2$.

It is important to note that $(f \circ g)(x)$ is not necessarily the same as $(g \circ f)(x)$. The process is not always commutative like addition or multiplication expressions. It *can* be commutative, but most often this is not the case.

Evaluating Linear Functions

A **function** can be expressed as an equation that relates an input to an output where each input corresponds to exactly one output. The input of a function is defined by the x-variable, and the output is defined by the y-variable. For example, consider the function $y = 2x + 6$. The value of y, the output, is determined by the value of the x, the input. If the value of x is 3, the value of y is $y = 2(3) + 6 = 6 + 6 = 12$. This means that when $x = 3$, $y = 12$. This can be expressed as the ordered pair (3,12).

It is common for function equations to use the form $f(x) =$ instead of $y =$. However, $f(x)$ and y represent the same thing. We read $f(x)$ as "f of x." The expression "f of x" implies that the value of f depends on the value of x. The function used in the example above could be expressed as $y = 2x + 6$ or $f(x) = 2x + 6$. Both functions represent the same line when graphed.

Functions that are expressed in the form $f(x) =$ are evaluated in the same way the equations are evaluated in the form $y =$. For example, when evaluating the function $f(x) = 3x - 2$ for $f(6)$, substitute 6 in for x, and simplify. In this case, $f(x) = 3x - 2$ becomes $f(6) = 3(6) - 2 = 18 - 2 = 16$. When x is 6, $f(x)$ is 16.

Example: To find the value of $f(8)$, calculate as follows:

$$f(x) = 3x - 2$$
$$f(8) = 3(8) - 2$$
$$f(8) = 22$$

Review Video: Evaluating Functions
Visit mometrix.com/academy and enter code: 588515

ADVANCED FUNCTIONS

STEP FUNCTIONS

The double brackets indicate a step function. For a step function, the value inside the double brackets is rounded down to the nearest integer. The graph of the function $f_0(x) = [\![x]\!]$ appears on the left graph. In comparison $f(x) = 2\left[\!\left[\frac{1}{3}(x-1)\right]\!\right]$ is on the right graph. The coefficient of 2 shows that it's stretched vertically by a factor of 2 (so there's a vertical distance of 2 units between successive "steps"). The coefficient of $\frac{1}{3}$ in front of the x shows that it's stretched horizontally by a factor of 3 (so each "step" is three units long), and the $x - 1$ shows that it's displaced one unit to the right.

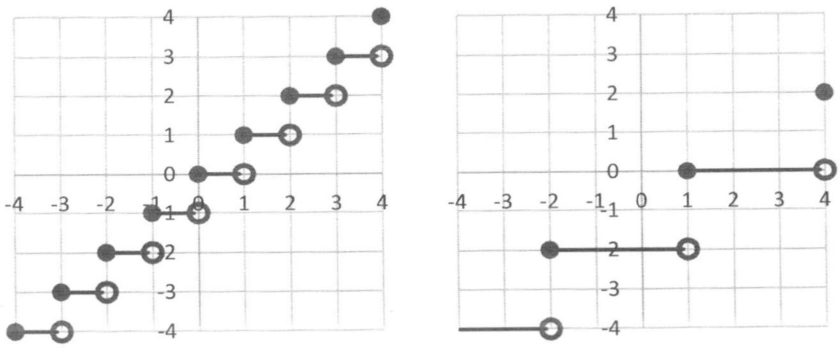

TRANSCENDENTAL FUNCTIONS

Roughly speaking, algebraic functions are functions defined by formulas involving numbers and variables combined by addition, subtraction, multiplication, division, powers (but not variable powers), and roots. **Transcendental functions** are functions that are not algebraic. A function that includes logarithms, trigonometric functions, or variables as exponents, is transcendental, not algebraic, even if the function also includes polynomials or roots.

EXPONENTIAL FUNCTIONS

Exponential functions are functions that have the form $y = b^x$, where base $b > 0$ and $b \neq 1$. The exponential function can also be written $f(x) = b^x$. The following properties apply to exponential expressions:

Property	Description
$a^x a^y = a^{x+y}$	The product of exponentials with the same base equals the base raised to the sum of the powers
$a^x / a^y = a^{x-y}$	The quotient of exponentials with the same base equals the base raised to the difference of the powers
$(a^x)^y = a^{xy}$	An exponential raised to a power equals the base raised to the product of the powers
$(ab)^x = a^x b^x$	Exponentiation distributes over multiplication
$(a/b)^x = a^x / b^x$	Exponentiation distributes over division

The graph of an example exponential function, $f(x) = 2^x$, is below:

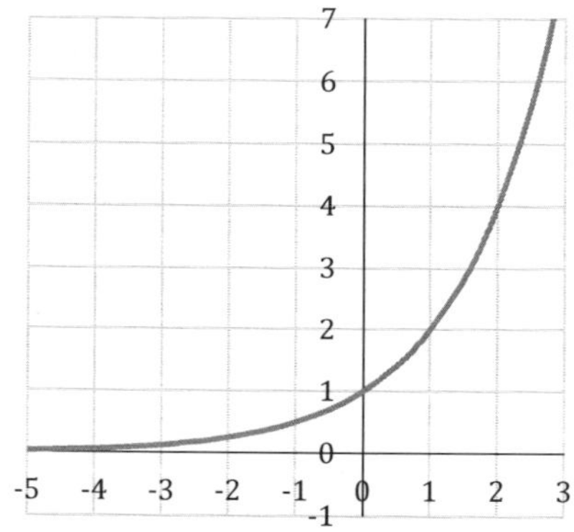

Note in the graph that the y-value approaches zero to the left and infinity to the right. One of the key features of an exponential function is that there will be one end that goes off to infinity and another that asymptotically approaches a lower bound. Common forms of exponential functions include the following:

Geometric sequences: The terms of a geometric sequence have the form $a_n = a_1 \times r^{n-1}$, where a_n is the value of the nth term, a_1 is the initial term, and r is the common ratio between succesive terms. Note that $a_1 \times r^{1-1} = a_1 \times r^0 = a_1 \times 1 = a_1$

> **Review Video: Geometric Sequences**
> Visit mometrix.com/academy and enter code: 140779

General exponential growth or decay: The general exponential growth or decay function is $f(t) = a(1+r)^t$, where the value of $f(t)$ is the amount (of whatever quantity we are measuring) at time t, the constant a is the initial amount (the amount at time $t = 0$), and the constant r is the fixed rate of increase (if $r > 0$) or decrease (if $r < 0$) per unit time. For example, if we invest \$1000 at an interest rate of 6% interest compounded annually and we measure time in years, then our investment will grow according to an exponential model with $a = \$1000$ and $r = 6\% = 0.06$. Thus, after t years the value of our investment will be $f(t) = \$1000(1+0.06)^t$. After 5 years, for instance, the investment will grow to $f(5) = \$1000(1+0.06)^5 \approx \$1338.23$.

Compound interest: We can modify the general exponential growth function above slightly to get a formula for interest compounded n times per year. Replacing the constant a with P (for principal, the initial amount we invest), the total value of our investment after t years is $f(t) = P\left(1+\frac{r}{n}\right)^{nt}$, where r is the nominal annual interest rate. Thus, if we change our \$1000 investment above by having the interest rate be 6% compounded *semiannually* (two times per year—so $n = 2$), the value of our investment after t years is $f(t) = \$1000\left(1+\frac{0.06}{2}\right)^{2t} = \$1000(1+0.03)^{2t}$. After 5 years, for instance, the investment will grow to $f(5) = \$1000(1+0.03)^{10} \approx \$1343.92$. This is slightly higher than the previous investment because the interest compounds twice as often.

> **Review Video: Compound Interest Formula**
> Visit mometrix.com/academy and enter code: 100366
>
> **Review Video: Interest Functions**
> Visit mometrix.com/academy and enter code: 559176

Population growth and continuously compounded interest: When a quantity grows (or decays) continuously at a rate that stays constant relative to the size the quantity has already attained, then we can model its size at time t with the function $f(t) = ae^{rt}$, where a is the initial amount, r is the relative growth rate (which we often call simply the **growth rate**), and e is the irrational constant known as Euler's number (approximately 2.718; scientific calculators have a button for finding e^x, which is known as the **natural exponential function**). For instance, under some circumstances, if an initial population (of people, plants, or animals, for example) of size a grows at constant relative rate r, then the size of the population at time t is $f(t) = ae^{rt}$. Similarly, if we invest principal P (instead of a) for t years at nominal annual interest rate r compounded *continuously*, then after t years the value of the investment is $f(t) = Pe^{rt}$.

For example, suppose the initial population of a town is $a = 1{,}200$ people and the relative annual growth rate is $r = 5\% = 0.05$. Then the population of the town after t years is $f(t) = ae^{rt} = 1200e^{0.05t}$. After 10 years, for instance, the town population is $f(10) = 1200e^{0.05 \cdot 10} = 1200e^{0.5} \approx$ 1978 people.

Review Video: Population Growth
Visit mometrix.com/academy and enter code: 109278

Logarithmic Functions

The **logarithmic function base b** is the function $y = \log_b x$ or $f(x) = \log_b x$, where the base b may be any positive number except one. The most common bases for logarithms are base 10 (the **common logarithm**) and base e (the **natural logarithm**). We usually write the common logarithm as $y = \log x$ (that is, $\log x$, with no base listed, means $\log_{10} x$) and the natural logarithm as $y = \ln x$ (that is, $\ln x$ means $\log_e x$).

Exponential functions and logarithmic functions with the same base are inverse functions. That is, if $f(x) = b^x$, then $f^{-1}(x) = \log_b x$. This means that the two equations $y = b^x$ (exponential form) and $x = \log_b y$ (logarithmic form) express the same relationship between the quantities x and y. We often solve problems involving logarithms by rewriting them in exponential form, and vice versa. Also, because of this inverse relationship, logarithms and exponentials cancel each other. That is, $\log_b b^x = x$ and $b^{\log_b x} = x$.

The following properties apply to logarithmic expressions:

Property	Description
$\log_b 1 = 0$	The log of 1 is equal to 0 for any base
$\log_b b = 1$	The log of the base is equal to 1
$\log_b b^p = p$	The log of the base raised to a power is equal to that power
$\log_b MN = \log_b M + \log_b N$	The log of a product is the sum of the log of each factor
$\log_b \frac{M}{N} = \log_b M - \log_b N$	The log of a quotient is equal to the log of the dividend minus the log of the divisor
$\log_b M^p = p \log_b M$	The log of a value raised to a power is equal to the power times the log of the value

Logarithms are helpful in solving equations in which the variable appears in an exponent. For instance, consider the example above in which we model the population of a town by the function $f(t) = 1200e^{0.05t}$. Suppose we want to know how long it will take the population of the town to double to 2400 people from its original value of 1200. We find this by solving the equation $1200e^{0.05t} = 2400$. First, we divide both sides of the equation by 1200 to get $e^{0.05t} = 2$. Then we take natural logarithms of both sides to get $\ln e^{0.05t} = \ln 2$, which simplifies (because natural logarithms and natural exponentials are inverse functions) to $0.05t = \ln 2$. Finally, we divide both sides by 0.05 to get $t = (\ln 2)/0.05$, which we evaluate with a calculator to get $t \approx 13.9$ years.

The graph of an example logarithmic function, $f(x) = \log_2(x + 2)$, is below:

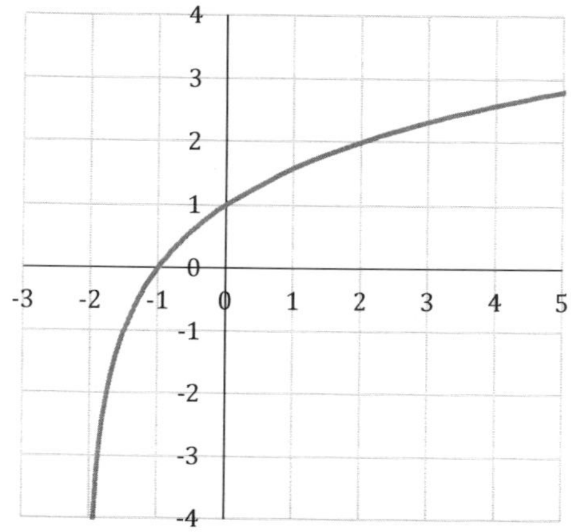

Review Video: Logarithmic Function
Visit mometrix.com/academy and enter code: 658985

Chapter Quiz

Ready to see how well you retained what you just read? Scan the QR code to go directly to the chapter quiz interface for this study guide. If you're using a computer, simply visit the online resources page at **mometrix.com/resources719/orelamgmath** and click the Chapter Quizzes link.

Measurement and Geometry

Measurement

Rounding and Estimation

Rounding is reducing the digits in a number while still trying to keep the value similar. The result will be less accurate but in a simpler form and easier to use. Whole numbers can be rounded to the nearest ten, hundred, or thousand, for instance.

To round a number, we make it a little smaller (rounding down) or a little larger (rounding up) to get a number that ends in zeros. We specify the number of zeros by naming the last place that we will not "zero out." For example, to round 8,3̲27 to the nearest hundred, we round down to 8,3̲00, zeroing out every digit to the right of the hundreds place. To round 4̲,728 to the nearest thousand, we round up to 5̲,000, increasing the thousands digit by one (to make the number larger) and zeroing out every digit to the right of the thousands place.

We decide whether to round down or up by looking at the first digit we are going to zero out. If it is less than 5 (namely, 0, 1, 2, 3, or 4) we round down. If it is greater than or equal to 5 (namely, 5, 6, 7, 8, or 9) we round up by adding 1 to the place we are rounding to. So, rounding 8,3̲27 to the nearest hundred, we round down to 8,3̲00 because the tens digit, 2, is less than 5. And rounding 4̲,728 to the nearest thousand, we round up to 5̲,000, increasing the thousands digit by 1, because the hundreds digit, 7, is greater than or equal to 5.

This even works with decimals. For example, rounding 39.7̲426 to the nearest tenth, we round down to 39.7̲000 (or simply 39.7̲) because the hundredths digit, 4, is less than 5. And rounding 0.019̲823 to the nearest thousandth, we round up to 0.020̲000 (or simply 0.02) by increasing the thousandths digit by 1, because the ten-thousandths digit, 8, is greater than or equal to 5.

When you are asked to estimate the solution to a problem, you will need to provide only an approximate figure or **estimation** for your answer. In this situation, you will need to round each number in the calculation to the level indicated (nearest hundred, nearest thousand, etc.) or to a level that makes sense for the numbers involved. When estimating a sum **all numbers must be rounded to the same level**. You cannot round one number to the nearest thousand while rounding another to the nearest hundred.

For instance, suppose you are considering buying four pieces of equipment for your home office. Their prices are \$485, \$1,217, \$750, and \$643. To estimate their total cost, you might round each price to the nearest hundred and add the rounded figures, getting an estimate of $\$500 + \$1{,}200 + \$800 + \$600 = \$3{,}100$. By estimating instead of making an exact calculation, you give up a little accuracy to get a simpler calculation.

Review Video: Rounding and Estimation
Visit mometrix.com/academy and enter code: 126243

Scientific Notation

Scientific notation is a way of writing large numbers in a shorter form. The form $a \times 10^n$ is used in scientific notation, where a is greater than or equal to 1 but less than 10, and n is the number of places the decimal must move to get from the original number to a. Example: The number 230,400,000 is cumbersome to write. To write the value in scientific notation, place a decimal point between the first and second numbers, and include all digits through the last non-zero digit ($a = 2.304$). To find the appropriate power of 10, count the number of places the decimal point had to move ($n = 8$). The number is positive if the decimal moved to the left, and negative if it moved to the right. We can then write 230,400,000 as 2.304×10^8. If we look instead at the number 0.00002304, we have the same value for a, but this time the decimal moved 5 places to the right ($n = -5$). Thus, 0.00002304 can be written as 2.304×10^{-5}. Using this notation makes it simple to compare very large or very small numbers. By comparing exponents, it is easy to see that 3.28×10^4 is smaller than 1.51×10^5, because 4 is less than 5.

Review Video: Scientific Notation
Visit mometrix.com/academy and enter code: 976454

Metric and Customary Measurements

Metric Measurement Prefixes

Giga-	One billion	1 *giga*watt is one billion watts
Mega-	One million	1 *mega*hertz is one million hertz
Kilo-	One thousand	1 *kilo*gram is one thousand grams
Deci-	One-tenth	1 *deci*meter is one-tenth of a meter
Centi-	One-hundredth	1 *centi*meter is one-hundredth of a meter
Milli-	One-thousandth	1 *milli*liter is one-thousandth of a liter
Micro-	One-millionth	1 *micro*gram is one-millionth of a gram

Review Video: How the Metric System Works
Visit mometrix.com/academy and enter code: 163709

Measurement Conversion

When converting between units, the goal is to maintain the same meaning but change the way it is displayed. In order to go from a larger unit to a smaller unit, multiply the number of the known amount by the equivalent amount. When going from a smaller unit to a larger unit, divide the number of the known amount by the equivalent amount.

For complicated conversions, it may be helpful to set up conversion fractions. In these fractions, one fraction is the **conversion factor**. The other fraction has the unknown amount in the numerator. So, the known value is placed in the denominator. Sometimes, the second fraction has the known value from the problem in the numerator and the unknown in the denominator. Multiply the two fractions to get the converted measurement. Note that since the numerator and the denominator of the factor are equivalent, the value of the fraction is 1. That is why we can say that the result in the new units is equal to the result in the old units even though they have different numbers.

It can often be necessary to chain known conversion factors together. As an example, consider converting 512 square inches to square meters. We know that there are 2.54 centimeters in an inch and 100 centimeters in a meter, and we know we will need to square each of these factors to achieve the conversion we are looking for.

$$\frac{512 \text{ in}^2}{1} \times \left(\frac{2.54 \text{ cm}}{1 \text{ in}}\right)^2 \times \left(\frac{1 \text{ m}}{100 \text{ cm}}\right)^2 = \frac{512 \cancel{\text{in}^2}}{1} \times \left(\frac{6.4516 \cancel{\text{cm}^2}}{1 \cancel{\text{in}^2}}\right) \times \left(\frac{1 \text{ m}^2}{10{,}000 \cancel{\text{cm}^2}}\right) = 0.330 \text{ m}^2$$

Review Video: Measurement Conversions
Visit mometrix.com/academy and enter code: 316703

Review Video: Converting Kilograms to Pounds
Visit mometrix.com/academy and enter code: 241463

Common Units and Equivalents

Metric Equivalents

1000 µg (microgram)	1 mg
1000 mg (milligram)	1 g
1000 g (gram)	1 kg
1000 kg (kilogram)	1 metric ton
1000 mL (milliliter)	1 L
1000 µm (micrometer)	1 mm
1000 mm (millimeter)	1 m
100 cm (centimeter)	1 m
1000 m (meter)	1 km

Distance and Area Measurement

Unit	Abbreviation	US equivalent	Metric equivalent
Inch	in	1 inch	2.54 centimeters
Foot	ft	12 inches	0.305 meters
Yard	yd	3 feet	0.914 meters
Mile	mi	5280 feet	1.609 kilometers
Acre	ac	4840 square yards	0.405 hectares
Square Mile	sq. mi. or mi.^{2}	640 acres	2.590 square kilometers

Capacity Measurements

Unit	Abbreviation	US equivalent	Metric equivalent
Fluid Ounce	fl oz	8 fluid drams	29.573 milliliters
Cup	c	8 fluid ounces	0.237 liter
Pint	pt.	16 fluid ounces	0.473 liter
Quart	qt.	2 pints	0.946 liter
Gallon	gal.	4 quarts	3.785 liters
Teaspoon	t or tsp.	1 fluid dram	5 milliliters
Tablespoon	T or tbsp.	4 fluid drams	15 or 16 milliliters
Cubic Centimeter	cc or cm^3	0.271 drams	1 milliliter

Weight Measurements

Unit	Abbreviation	US equivalent	Metric equivalent
Ounce	oz	16 drams	28.35 grams
Pound	lb	16 ounces	453.6 grams
Ton	tn.	2,000 pounds	907.2 kilograms

Volume and Weight Measurement Clarifications

Always be careful when using ounces and fluid ounces. They are not equivalent.

1 pint = 16 fluid ounces	1 fluid ounce ≠ 1 ounce
1 pound = 16 ounces	1 pint ≠ 1 pound

Having one pint of something does not mean you have one pound of it. In the same way, just because something weighs one pound does not mean that its volume is one pint.

In the United States, the word "ton" by itself refers to a short ton or a net ton. Do not confuse this with a long ton (also called a gross ton) or a metric ton (also spelled *tonne*), which have different measurement equivalents.

$$1 \text{ US ton} = 2000 \text{ pounds} \quad \neq \quad 1 \text{ metric ton} = 1000 \text{ kilograms}$$

Precision, Accuracy, and Error

Measurements of physical quantities (e.g., length, area, volume, weight, mass, and time) in the real world are never perfect. Measurements miss the true value, and repeated measurements produce different values. For this reason, fields that depend on good measurement have technical terms, precision and accuracy, that describe how particular ways of measuring a quantity produce good or bad results. Note that these terms apply not to single measurements but to repeated measurements of the same quantity using the same procedure.

Precision describes the consistency of measurements. Measurements that cluster closely together, with little variation from one measurement to the next, have high precision. Measurements that vary a great deal from one measurement to the next have low precision.

Accuracy describes the closeness of measurements to the true value of the quantity being measured or, more technically, the tendency of measurements to cluster around the true value (though we do not usually know the true value; otherwise, we would not bother to measure it).

An analogy from archery may help. An outstanding archer aiming at the center of the target (the true value) produces a tight cluster of arrows around the center (high accuracy, high precision). Under the same circumstances a modestly good archer produces a loose cluster around the center (high accuracy, low precision). If, however, there is a strong crosswind and the archers do nothing to compensate for it, the outstanding archer's shots will cluster tightly around a point away from the center (low accuracy, high precision); and the modestly good archer's shots will cluster loosely around a point away from the center (low accuracy, low precision).

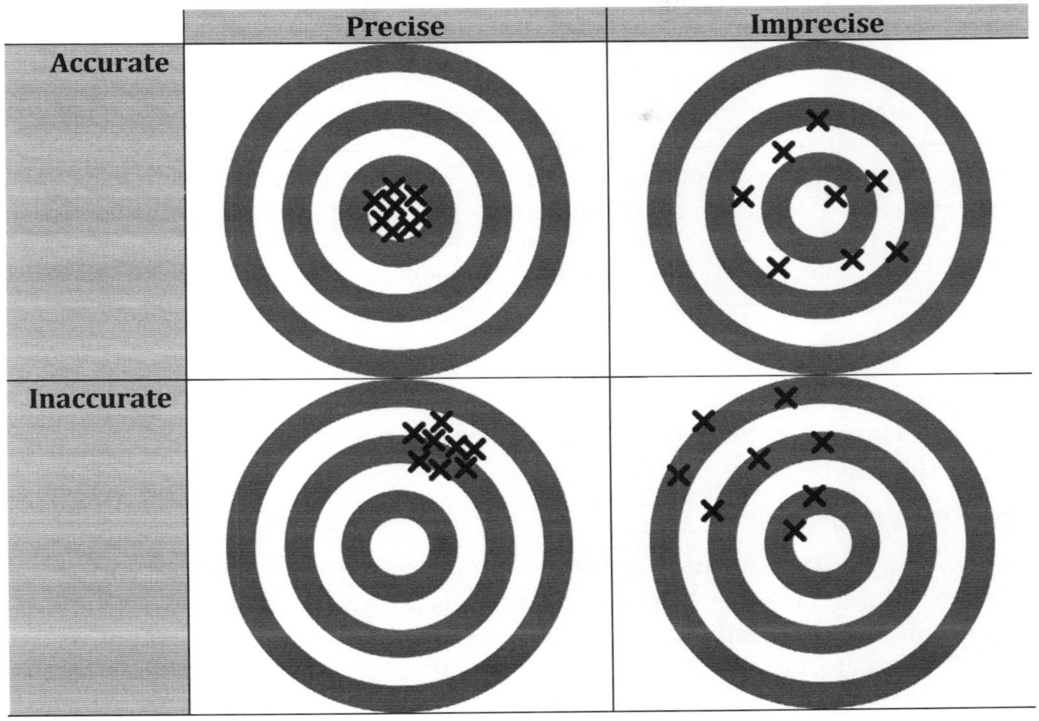

Error in measurement is not a mistake but simply the difference between the true value and the measured value. Since we seldom know the true value, we seldom know the exact error; but sometimes we know a limit on how big the error can be (an error bound), in which case we can use the plus-or-minus sign, $\pm$, to indicate how far the true value might lie from the measured value. For instance, if we measure the length of a metal bar as 25.6 cm and we know the error in our measurement is no more than 0.03 cm (some sources call this **approximate error**), then we might report the length of the bar as 25.6 ± 0.03 cm, which means it lies between 25.57 cm and 25.63 cm.

A particularly important application of this comes from the necessity of rounding whenever we make a measurement. For instance, if we measure the length of a pencil using a ruler marked off in whole inches (no fractions), then we must round our measurement to the nearest inch. Thus, if we report the length as 8 inches, it means that the actual value lies somewhere between 7.5 inches and 8.5 inches, which is to say 8 ± 0.5 inches. This built-in error bound due to rounding is called the **maximum possible error**, and it is always half the magnitude of the smallest unit used in the measurement. Determining the smallest unit can be somewhat subtle, involving the rules for significant figures commonly used in physics, chemistry, and other fields.

Review Video: Precision, Accuracy, and Error
Visit mometrix.com/academy and enter code: 520377

Points, Lines, and Planes

Points, Lines, and Planes

Points and Lines

A **point** is a fixed location in space, has no size or dimensions, and is commonly represented by a dot. A **line** is a set of points that extends infinitely in two opposite directions. It has length, but no width or depth. A line can be defined by any two distinct points that it contains. A **line segment** is a portion of a line that has definite endpoints. A **ray** is a portion of a line that extends from a single point on that line in one direction along the line. It has a definite beginning, but no ending.

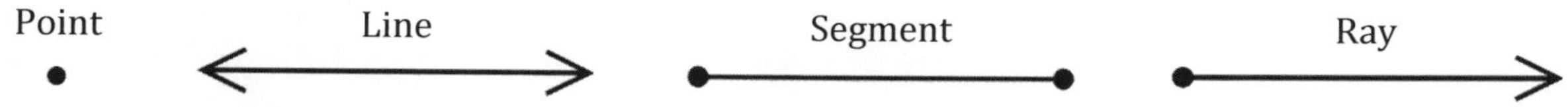

Points are **collinear** if there is a single line that passes through all of them. Otherwise, they are noncollinear. Two points are always collinear since two points define a line. Three points may be noncollinear. For example, the three vertices of a triangle are noncollinear since there is no line that goes through all three of them.

Interactions Between Lines

Intersecting lines are lines that have exactly one point in common. **Concurrent lines** are multiple lines that intersect at a single point. **Perpendicular lines** are lines that intersect at right angles. They are represented by the symbol ⊥. The shortest distance from a line to a point not on the line is a perpendicular segment from the point to the line. **Parallel lines** are lines in the same plane that have no points in common and never meet. Two distinct lines in a given plane are always either intersecting or parallel. **Skew lines** are two distinct lines in a three dimensional space that do not intersect and may also not be parallel because there is no single plane that contains them both.

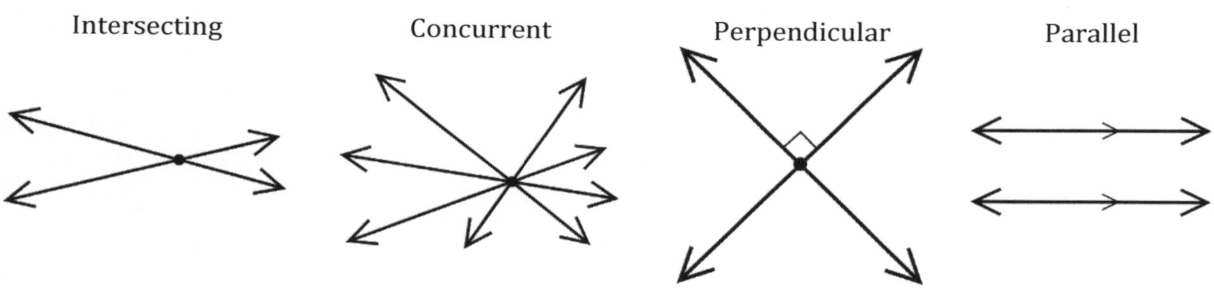

Review Video: Parallel and Perpendicular Lines
Visit mometrix.com/academy and enter code: 815923

A **transversal** is a line that intersects at least two other lines, which may or may not be parallel to one another. A transversal that intersects parallel lines is a common occurrence in geometry. A **bisector** is a line or line segment that divides another line segment into two equal lengths. A

perpendicular bisector of a line segment is composed of points that are equidistant from the endpoints of the segment it is dividing.

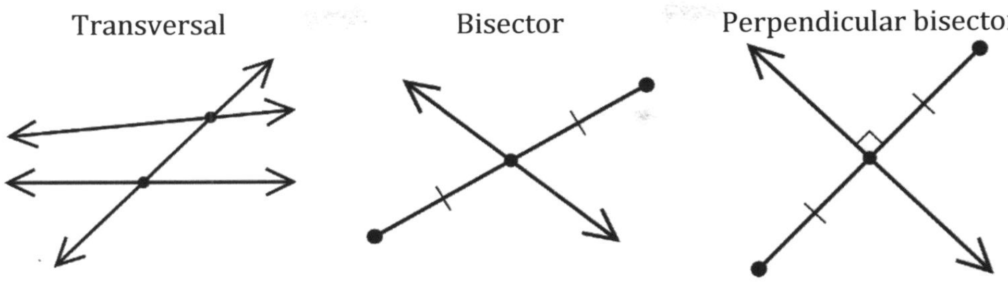

The **projection of a point on a line** is the point at which a perpendicular line drawn from the given point to the given line intersects the line. This is also the shortest distance from the given point to the line. The **projection of a segment on a line** is a segment whose endpoints are the points formed when perpendicular lines are drawn from the endpoints of the given segment to the given line. This is similar to the length a diagonal line appears to be when viewed from above.

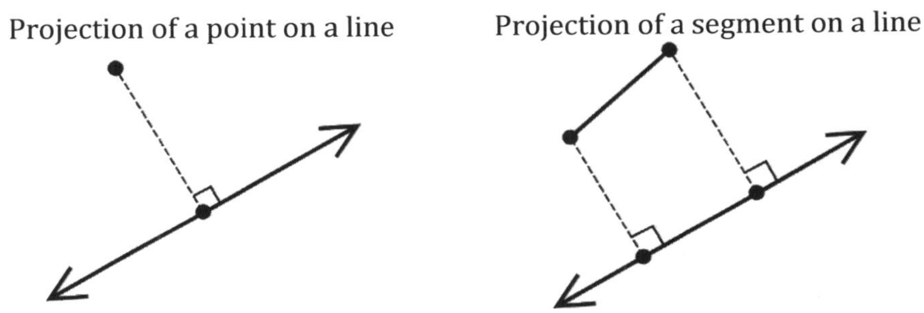

PLANES

A **plane** is a two-dimensional flat surface defined by three non-collinear points. A plane extends an infinite distance in all directions in those two dimensions. It contains an infinite number of points, parallel lines and segments, intersecting lines and segments, as well as parallel or intersecting rays. A plane will never contain a three-dimensional figure or skew lines. Two given planes are either parallel or they intersect at a line. A plane may intersect a circular conic surface to form **conic sections**, such as a parabola, hyperbola, circle or ellipse.

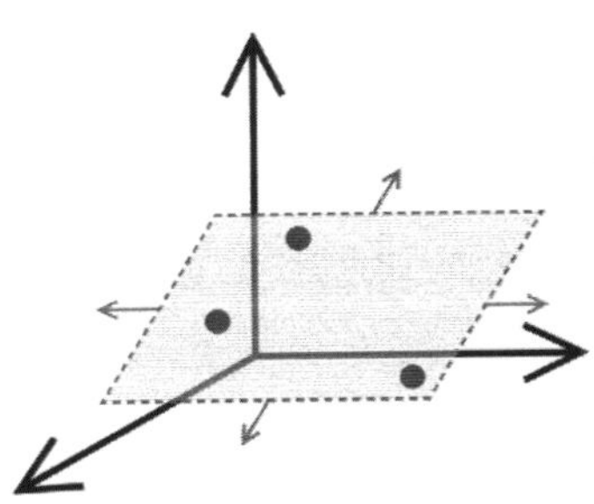

Review Video: Lines and Planes
Visit mometrix.com/academy and enter code: 554267

Angles

ANGLES

ANGLES AND VERTICES

An **angle** is formed when two lines or line segments meet at a common point. It may be a common starting point for a pair of segments or rays, or it may be the intersection of lines. Angles are represented by the symbol ∠.

The **vertex** is the point at which two segments or rays meet to form an angle. If the angle is formed by intersecting rays, lines, and/or line segments, the vertex is the point at which four angles are formed. The pairs of angles opposite one another are called vertical angles, and their measures are equal.

- An **acute** angle is an angle with a degree measure less than 90°.
- A **right** angle is an angle with a degree measure of exactly 90°.
- An **obtuse** angle is an angle with a degree measure greater than 90° but less than 180°.
- A **straight angle** is an angle with a degree measure of exactly 180°.
- A **reflex angle** is an angle with a degree measure greater than 180° but less than 360°.
- A **full angle** is an angle with a degree measure of exactly 360°.

Review Video: Angles
Visit mometrix.com/academy and enter code: 264624

RELATIONSHIPS BETWEEN ANGLES

Two angles whose sum is exactly 90° are said to be **complementary**. The two angles may or may not be adjacent. In a right triangle, the two acute angles are complementary.

Two angles whose sum is exactly 180° are said to be **supplementary**. The two angles may or may not be adjacent. Two intersecting lines always form two pairs of supplementary angles. Adjacent supplementary angles will always form a straight line.

Two angles that have the same vertex and share a side are said to be **adjacent**. Vertical angles are not adjacent because they share a vertex but no common side.

Review Video: Adjacent Angles
Visit mometrix.com/academy and enter code: 100375

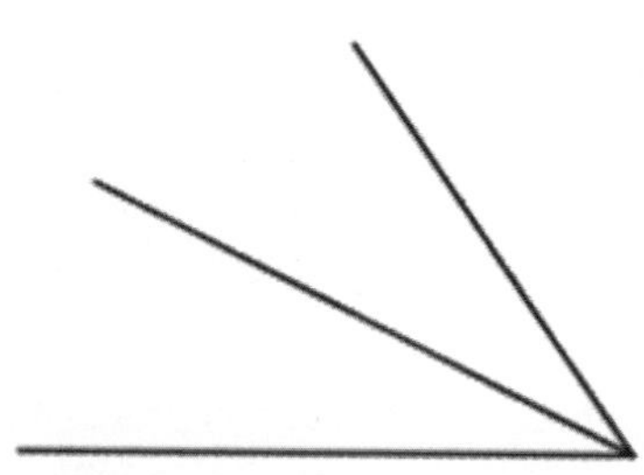

Adjacent
Share vertex and side

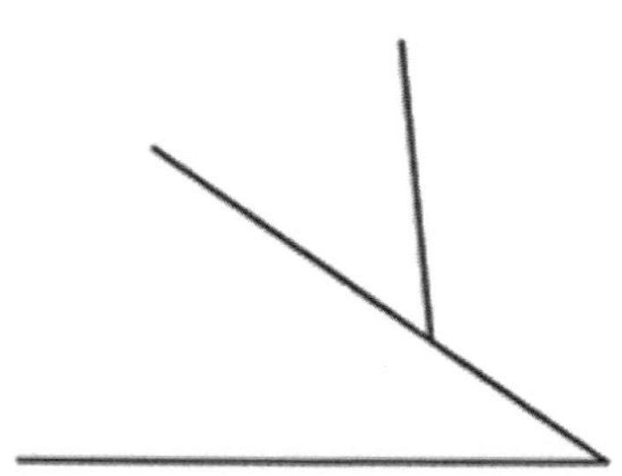

Not adjacent
Share part of a side, but not vertex

When two lines are cut by a transversal, the angles that are between the two lines are **interior angles**. In the diagram below, angles 3, 4, 5, and 6 are interior angles.

When two lines are cut by a transversal, the angles that are outside the lines are **exterior angles**. In the diagram below, angles 1, 2, 7, and 8 are exterior angles.

When two lines are cut by a transversal, the angles that are in the same position relative to the transversal and the cut lines are **corresponding angles**. The diagram below has four pairs of corresponding angles: angles 1 and 5, angles 2 and 6, angles 3 and 7, and angles 4 and 8. Corresponding angles formed by parallel lines are congruent.

When two lines are cut by a transversal, the two interior angles that are on opposite sides of the transversal are called **alternate interior angles**. In the diagram below, there are two pairs of alternate interior angles: angles 3 and 6, and angles 4 and 5. Alternate interior angles formed by parallel lines are congruent. Similarly, the two interior angles on the same side of the transversal (angles 3 and 5, and angles 4 and 6) are supplementary when the transversed lines are parallel. Some books call these angles **same side interior angles**.

When two lines are cut by a transversal, the two exterior angles that are on opposite sides of the transversal are called **alternate exterior angles**. In the diagram below, there are two pairs of alternate exterior angles: angles 1 and 8, and angles 2 and 7. Alternate exterior angles formed by parallel lines are congruent. Similarly, the two exterior angles on the same side of the transversal (angles 1 and 7, and angles 2 and 8) are supplementary when the transversed lines are parallel. Some books call these angles **same side exterior angles**.

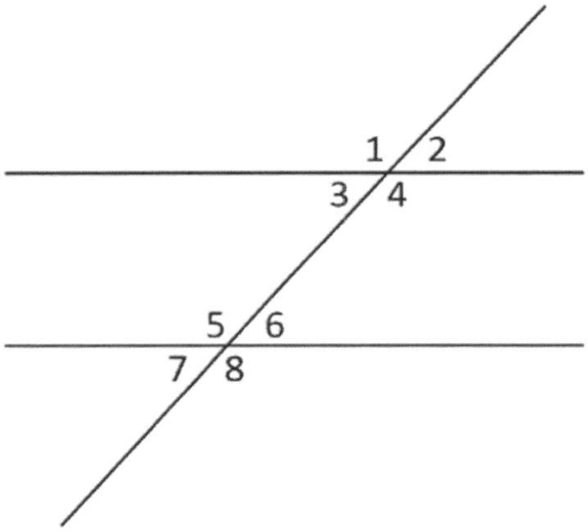

When two lines intersect, four angles are formed. The non-adjacent angles at this vertex are called vertical angles. Vertical angles are congruent. In the diagram, $\angle ABD \cong \angle CBE$ and $\angle ABC \cong \angle DBE$. The other pairs of angles, $(\angle ABC, \angle CBE)$ and $(\angle ABD, \angle DBE)$, are supplementary, meaning the pairs sum to 180°.

Review Video: Congruent Angles
Visit mometrix.com/academy and enter code: 642874

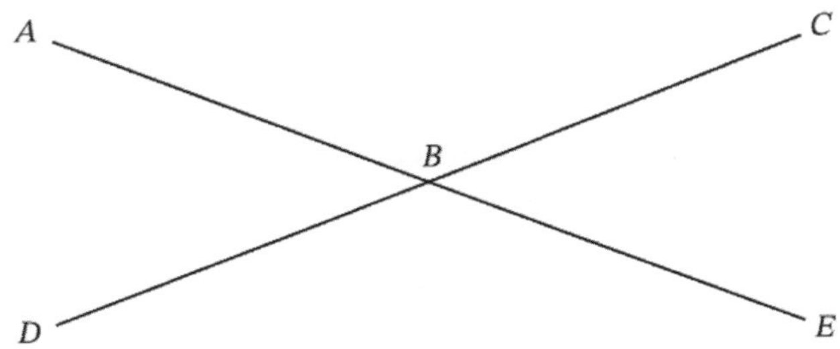

Polygons

Polygons

A **polygon** is a closed, two-dimensional figure with three or more straight line segments called **sides**. The point at which two sides of a polygon intersect is called the **vertex**. In a polygon, the number of sides is always equal to the number of vertices. A polygon with all sides congruent and all angles equal is called a **regular polygon**. Common polygons are:

Triangle = 3 sides
Quadrilateral = 4 sides
Pentagon = 5 sides
Hexagon = 6 sides
Heptagon = 7 sides
Octagon = 8 sides
Nonagon = 9 sides
Decagon = 10 sides
Dodecagon = 12 sides

More generally, an n-gon is a polygon that has n angles and n sides.

Review Video: Intro to Polygons
Visit mometrix.com/academy and enter code: 271869

The sum of the interior angles of an n-sided polygon is $(n - 2) \times 180°$. For example, in a triangle $n = 3$. So the sum of the interior angles is $(3 - 2) \times 180° = 180°$. In a quadrilateral, $n = 4$, and the sum of the angles is $(4 - 2) \times 180° = 360°$.

Review Video: Sum of Interior Angles
Visit mometrix.com/academy and enter code: 984991

Convex and Concave Polygons

A **convex polygon** is a polygon whose diagonals all lie within the interior of the polygon. A **concave polygon** is a polygon with at least one diagonal that is outside the polygon. In the diagram below, quadrilateral $ABCD$ is concave because diagonal $\overline{AC}$ lies outside the polygon and quadrilateral $EFGH$ is convex because both diagonals lie inside the polygon.

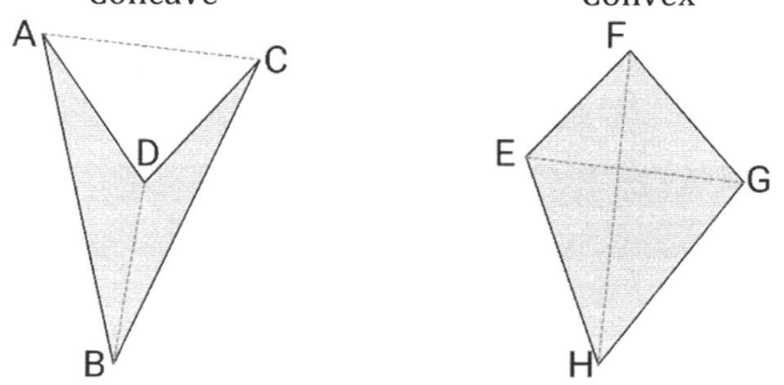

Apothem and Radius

A line segment from the center of a regular polygon that is perpendicular to a side of the polygon is called the **apothem**. A line segment from the center of a regular polygon to a vertex of the polygon is called a **radius**. In a regular polygon, the apothem can be used to find the area of the polygon using the formula $A = \frac{1}{2}ap$, where a is the apothem, and p is the perimeter.

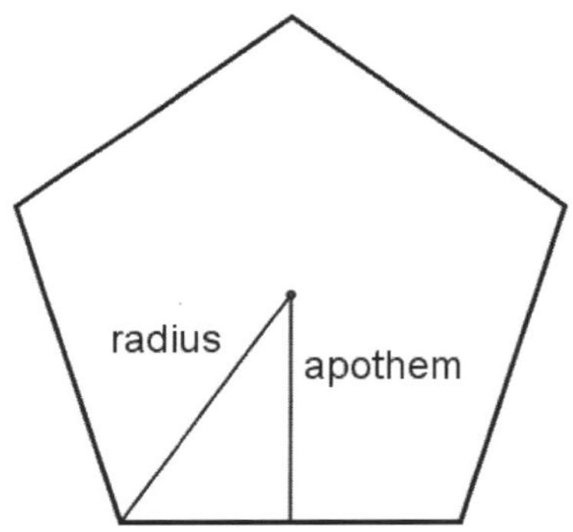

A **diagonal** is a line segment that joins two non-adjacent vertices of a polygon. The number of diagonals a polygon has can be found by using the formula:

$$\text{number of diagonals} = \frac{n(n-3)}{2}$$

Note that n is the number of sides in the polygon. This formula works for all polygons, not just regular polygons.

Congruence and Similarity

Congruent figures are geometric figures that have the same size and shape. For congruent polygons all corresponding angle measures are equal, and all corresponding side lengths are equal. Congruence is indicated by the symbol ≅. For instance, the expression $ABC \cong DEF$ indicates that the triangles below are congruent. The order of the letters is important, indicating which parts of the polygons correspond to each other. For example, since the letters A and D both come first, $\angle A$ and $\angle D$ have the same measure.

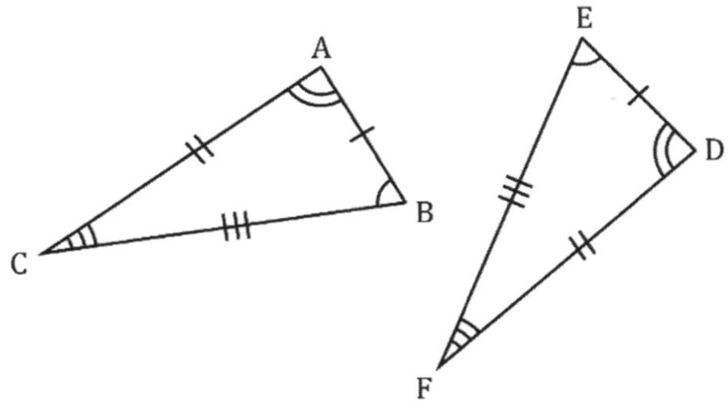

Similar figures are geometric figures that have the same shape, but do not necessarily have the same size. For similar polygons all corresponding angle measures are equal, and all corresponding side lengths are proportional, but they do not have to be equal. It is indicated by the symbol $\sim$. For instance, the expression $ABC \sim DEF$ indicates that the triangles below are similar. Again, the order of the letters indicates which parts of the polygons correspond to each other.

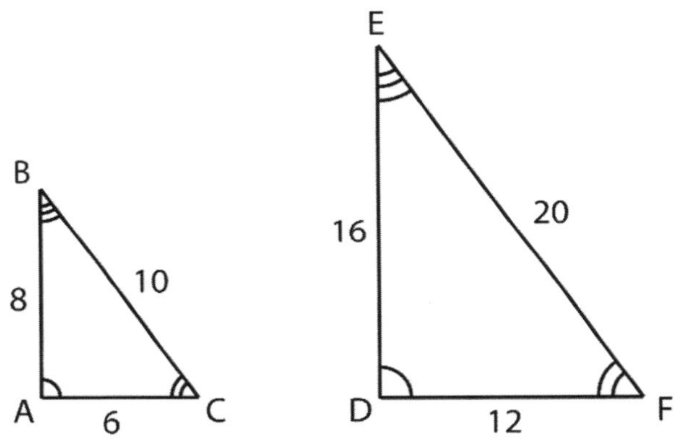

Note that all congruent figures are also similar, but not all similar figures are congruent.

Review Video: Congruent Shapes
Visit mometrix.com/academy and enter code: 492281

LINE OF SYMMETRY

A line that divides a figure or object into congruent parts that are mirror images of each other across the line is called a **line of symmetry**. An object may have no lines of symmetry, one line of symmetry, or multiple (i.e., more than one) lines of symmetry.

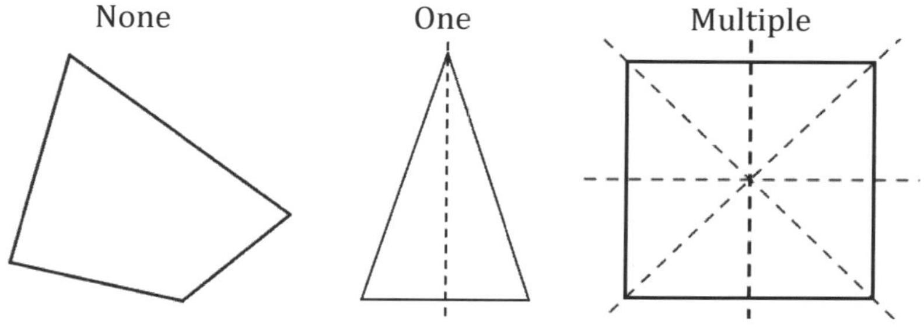

Review Video: Symmetry
Visit mometrix.com/academy and enter code: 528106

TRIANGLES

A triangle is a three-sided figure with the sum of its interior angles being 180°. The **perimeter of any triangle** is found by summing the three side lengths; $P = a + b + c$. For an equilateral triangle, this is the same as $P = 3a$, where a is any side length, since all three sides are the same length.

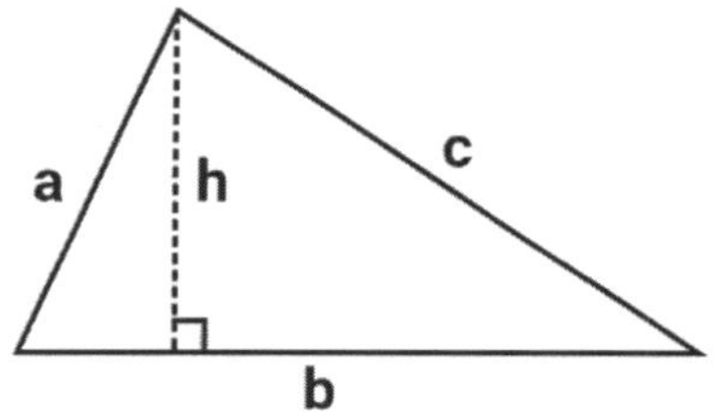

Review Video: Proof that a Triangle is 180 Degrees
Visit mometrix.com/academy and enter code: 687591

Review Video: Area and Perimeter of a Triangle
Visit mometrix.com/academy and enter code: 853779

The **area of any triangle** can be found by taking half the product of one side length referred to as the base, often given the variable b and the perpendicular distance from that side to the opposite vertex called the altitude or height and given the variable h. In equation form that is $A = \frac{1}{2}bh$. Another formula that works for any triangle is $A = \sqrt{s(s-a)(s-b)(s-c)}$, where s is the semiperimeter: $\frac{a+b+c}{2}$, and a, b, and c are the lengths of the three sides. Special cases include isosceles triangles, $A = \frac{1}{2}b\sqrt{a^2 - \frac{b^2}{4}}$, where b is the unique side and a is the length of one of the two congruent sides, and equilateral triangles, $A = \frac{\sqrt{3}}{4}a^2$, where a is the length of a side.

Review Video: Area of Any Triangle
Visit mometrix.com/academy and enter code: 138510

PARTS OF A TRIANGLE

An **altitude** of a triangle is a line segment drawn from one vertex perpendicular to the opposite side. In the diagram that follows, $\overline{BE}$, $\overline{AD}$, and $\overline{CF}$ are altitudes. The length of an altitude is also called the height of the triangle. The three altitudes in a triangle are always concurrent. The point of concurrency of the altitudes of a triangle, O, is called the **orthocenter**. Note that in an obtuse triangle, the orthocenter will be outside the triangle, and in a right triangle, the orthocenter is the vertex of the right angle.

A **median** of a triangle is a line segment drawn from one vertex to the midpoint of the opposite side. In the diagram that follows, $\overline{BH}$, $\overline{AG}$, and $\overline{CI}$ are medians. This is not the same as the altitude, except the altitude to the base of an isosceles triangle and all three altitudes of an equilateral triangle. The point of concurrency of the medians of a triangle, T, is called the **centroid**. This is the same point as the orthocenter only in an equilateral triangle. Unlike the orthocenter, the centroid is always inside the triangle. The centroid can also be considered the exact center of the triangle. Any shape triangle can be perfectly balanced on a tip placed at the centroid. The centroid is also the point that is two-thirds the distance from the vertex to the opposite side.

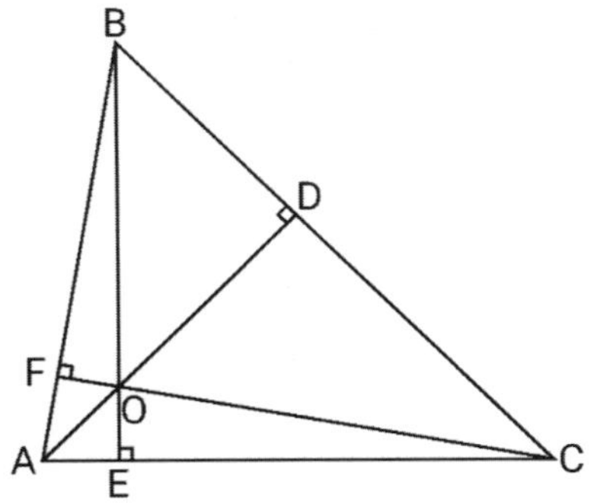

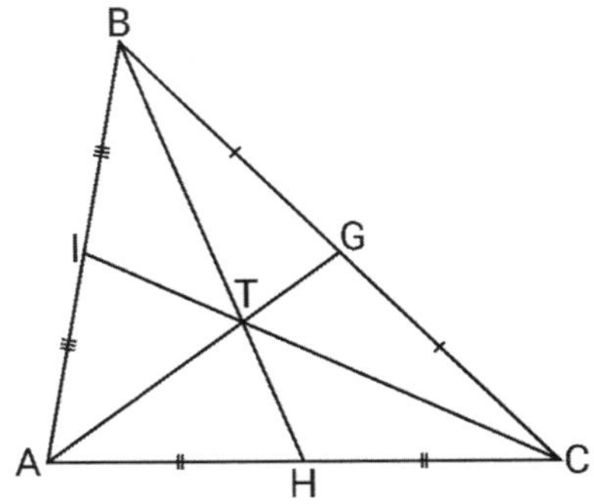

Review Video: Centroid, Incenter, Circumcenter, and Orthocenter
Visit mometrix.com/academy and enter code: 598260

TRIANGLE PROPERTIES

CLASSIFICATIONS OF TRIANGLES

A **scalene triangle** is a triangle with no congruent sides. A scalene triangle will also have three angles of different measures. The angle with the largest measure is opposite the longest side, and the angle with the smallest measure is opposite the shortest side. An **acute triangle** is a triangle whose three angles are all less than 90°. If two of the angles are equal, the acute triangle is also an **isosceles triangle**. An isosceles triangle will also have two congruent angles opposite the two congruent sides. If the three angles are all equal, the acute triangle is also an **equilateral triangle**. An equilateral triangle will also have three congruent angles, each 60°. All equilateral triangles are also acute triangles. An **obtuse triangle** is a triangle with exactly one angle greater than 90°. The other two angles may or may not be equal. If the two remaining angles are equal, the obtuse triangle is also an isosceles triangle. A **right triangle** is a triangle with exactly one angle equal to 90°. All right triangles follow the Pythagorean theorem. A right triangle can never be acute or obtuse.

The table below illustrates how each descriptor places a different restriction on the triangle:

Angles / Sides	**Acute: All angles < 90°**	**Obtuse: One angle > 90°**	**Right: One angle = 90°**
Scalene: No equal side lengths	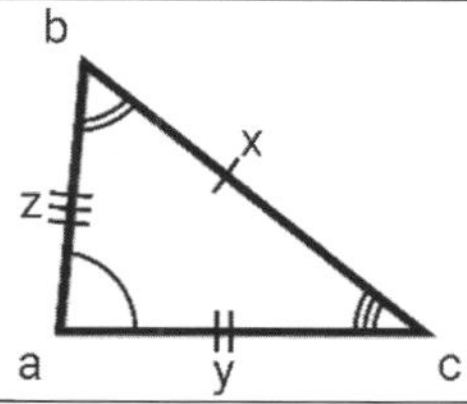	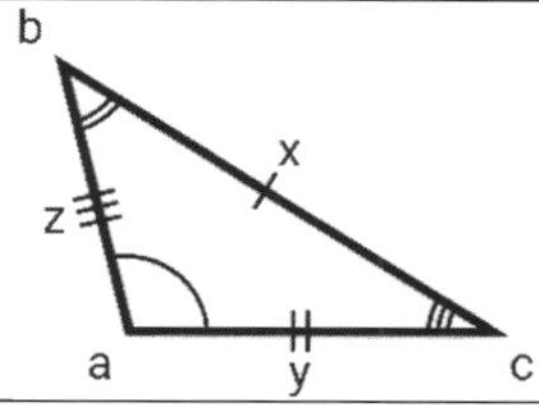	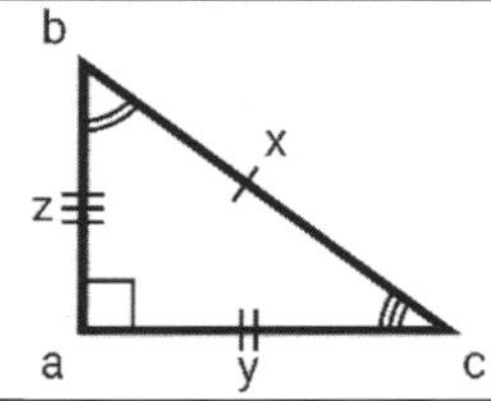
	$90° > \angle a > \angle b > \angle c$ $x > y > z$	$\angle a > 90° > \angle b > \angle c$ $x > y > z$	$90° = \angle a > \angle b > \angle c$ $x > y > z$
Isosceles: Two equal side lengths	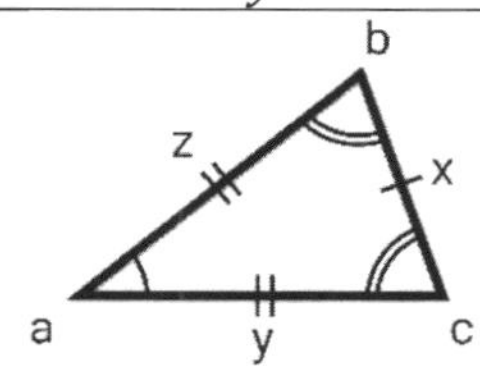	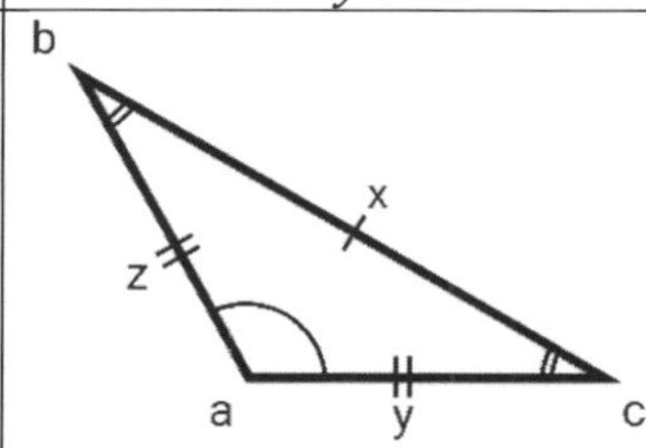	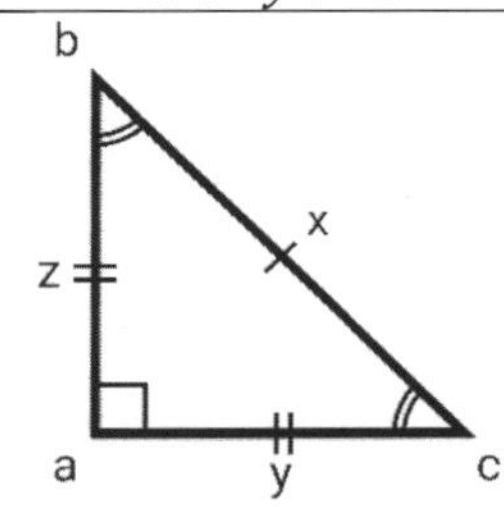
	$90° > \angle a, \angle b, or\ \angle c$ $\angle b = \angle c, \quad y = z$	$\angle a > 90° > \angle b = \angle c$ $x > y = z$	$\angle a = 90°$ $\angle b = \angle c = 45°$ $x > y = z$
Equilateral: Three equal side lengths	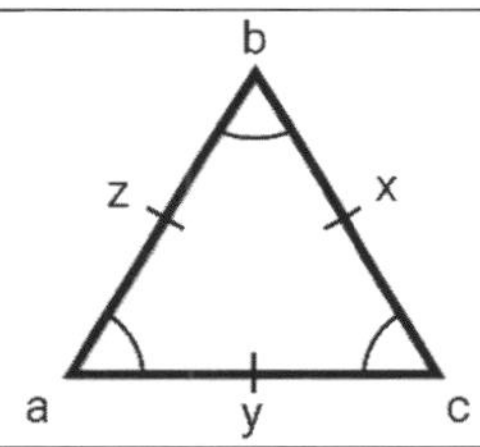		
	$60° = \angle a = \angle b = \angle c$ $x = y = z$		

Review Video: Introduction to Types of Triangles
Visit mometrix.com/academy and enter code: 511711

GENERAL RULES FOR TRIANGLES

The **triangle inequality theorem** states that the sum of the measures of any two sides of a triangle is always greater than the measure of the third side. If the sum of the measures of two sides were equal to the third side, a triangle would be impossible because the two sides would lie flat across the third side and there would be no vertex. If the sum of the measures of two of the sides was less than the third side, a closed figure would be impossible because the two shortest sides would never meet. In other words, for a triangle with sides lengths A, B, and C: $A + B > C, B + C > A$, and $A + C > B$.

The sum of the measures of the interior angles of a triangle is always 180°. Therefore, a triangle can never have more than one angle greater than or equal to 90°.

In any triangle, the angles opposite congruent sides are congruent, and the sides opposite congruent angles are congruent. The largest angle is always opposite the longest side, and the smallest angle is always opposite the shortest side.

The line segment that joins the midpoints of any two sides of a triangle is always parallel to the third side and exactly half the length of the third side.

Review Video: General Rules (Triangle Inequality Theorem)
Visit mometrix.com/academy and enter code: 166488

Similarity and Congruence Rules

Similar triangles are triangles whose corresponding angles are equal and whose corresponding sides are proportional. Represented by AAA. Similar triangles whose corresponding sides are congruent are also congruent triangles.

Triangles can be shown to be **congruent** in 5 ways:

- **SSS**: Three sides of one triangle are congruent to the three corresponding sides of the second triangle.
- **SAS**: Two sides and the included angle (the angle formed by those two sides) of one triangle are congruent to the corresponding two sides and included angle of the second triangle.
- **ASA**: Two angles and the included side (the side that joins the two angles) of one triangle are congruent to the corresponding two angles and included side of the second triangle.
- **AAS**: Two angles and a non-included side of one triangle are congruent to the corresponding two angles and non-included side of the second triangle.
- **HL**: The hypotenuse and leg of one right triangle are congruent to the corresponding hypotenuse and leg of the second right triangle.

Review Video: Similar Triangles
Visit mometrix.com/academy and enter code: 398538

Quadrilaterals

A **quadrilateral** is a closed two-dimensional geometric figure that has four straight sides. The sum of the interior angles of any quadrilateral is 360°.

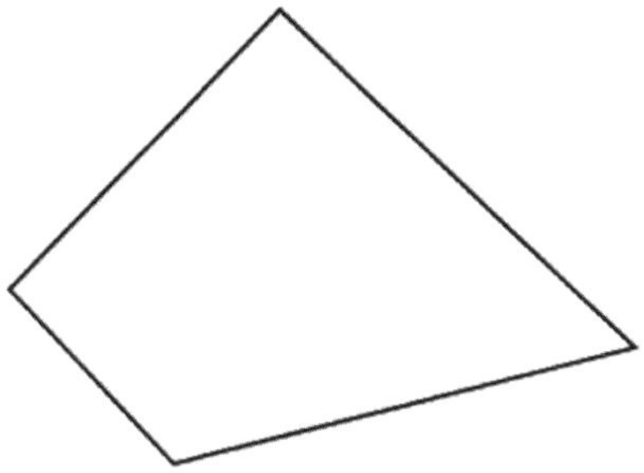

Review Video: Diagonals of Parallelograms, Rectangles, and Rhombi
Visit mometrix.com/academy and enter code: 320040

Kite

A **kite** is a quadrilateral with two pairs of adjacent sides that are congruent. A result of this is perpendicular diagonals. A kite can be concave or convex and has one line of symmetry.

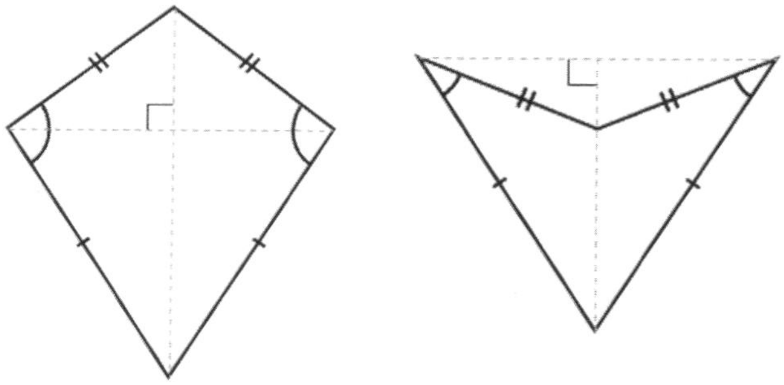

Trapezoid

Trapezoid: A trapezoid is defined as a quadrilateral that has at least one pair of parallel sides. There are no rules for the second pair of sides. So, there are no rules for the diagonals and no lines of symmetry for a trapezoid.

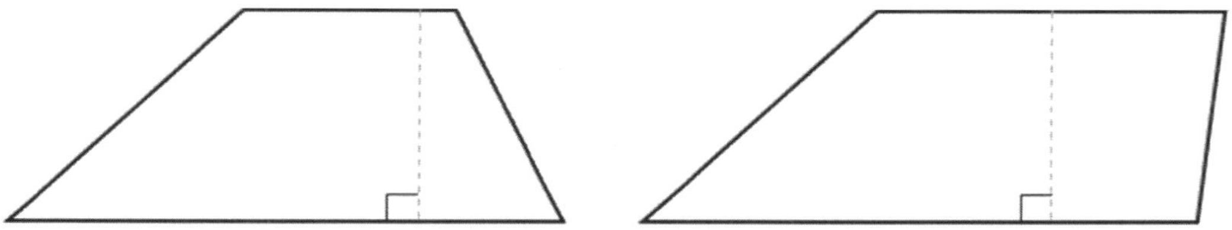

The **area of a trapezoid** is found by the formula $A = \frac{1}{2}h(b_1 + b_2)$, where h is the height (segment joining and perpendicular to the parallel bases), and b_1 and b_2 are the two parallel sides (bases). Do not use one of the other two sides as the height unless that side is also perpendicular to the parallel bases.

The **perimeter of a trapezoid** is found by the formula $P = a + b_1 + c + b_2$, where a, b_1, c, and b_2 are the four sides of the trapezoid.

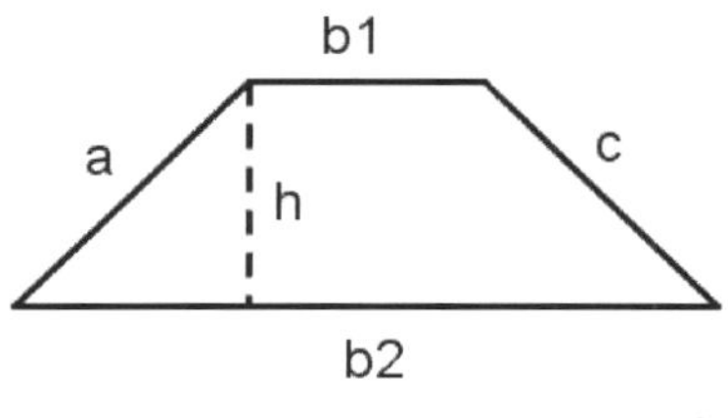

Review Video: Area and Perimeter of a Trapezoid
Visit mometrix.com/academy and enter code: 587523

Isosceles trapezoid: A trapezoid with equal base angles. This gives rise to other properties including: the two nonparallel sides have the same length, the two non-base angles are also equal, and there is one line of symmetry through the midpoints of the parallel sides.

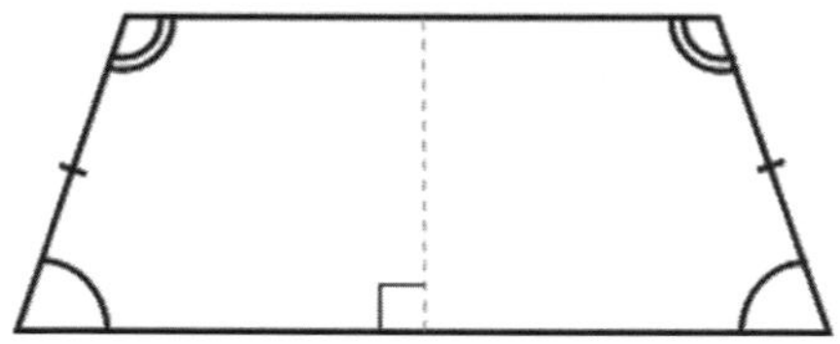

Parallelogram

A **parallelogram** is a quadrilateral that has two pairs of opposite parallel sides. As such it is a special type of trapezoid. The sides that are parallel are also congruent. The opposite interior angles are always congruent, and the consecutive interior angles are supplementary. The diagonals of a parallelogram divide each other. Each diagonal divides the parallelogram into two congruent triangles. A parallelogram has no line of symmetry, but does have 180-degree rotational symmetry about the midpoint.

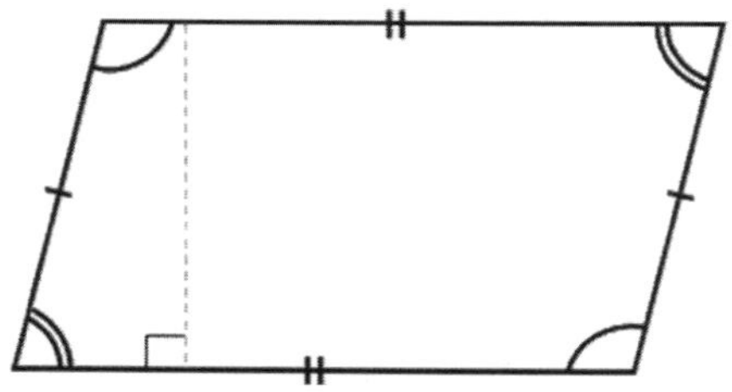

The **area of a parallelogram** is found by the formula $A = bh$, where b is the length of the base, and h is the height. Note that the base and height correspond to the length and width in a rectangle, so this formula would apply to rectangles as well. Do not confuse the height of a parallelogram with the length of the second side. The two are only the same measure in the case of a rectangle.

The **perimeter of a parallelogram** is found by the formula $P = 2a + 2b$ or $P = 2(a + b)$, where a and b are the lengths of the two sides.

Review Video: Area and Perimeter of a Parallelogram
Visit mometrix.com/academy and enter code: 718313

Rectangle

A **rectangle** is a quadrilateral with four right angles. All rectangles are parallelograms and trapezoids, but not all parallelograms or trapezoids are rectangles. The diagonals of a rectangle are congruent. Rectangles have two lines of symmetry (through each pair of opposing midpoints) and 180-degree rotational symmetry about the midpoint.

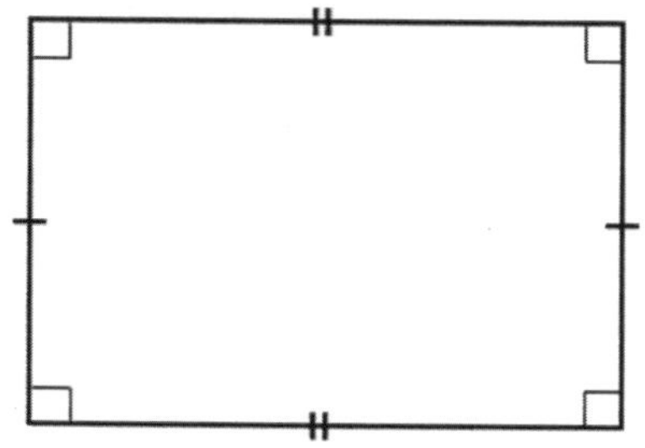

The **area of a rectangle** is found by the formula $A = lw$, where A is the area of the rectangle, l is the length (usually considered to be the longer side) and w is the width (usually considered to be the shorter side). The numbers for l and w are interchangeable.

The **perimeter of a rectangle** is found by the formula $P = 2l + 2w$ or $P = 2(l + w)$, where l is the length, and w is the width. It may be easier to add the length and width first and then double the result, as in the second formula.

Rhombus

A **rhombus** is a quadrilateral with four congruent sides. All rhombuses are parallelograms and kites; thus, they inherit all the properties of both types of quadrilaterals. The diagonals of a rhombus are perpendicular to each other. Rhombi have two lines of symmetry (along each of the diagonals) and 180° rotational symmetry. The **area of a rhombus** is half the product of the diagonals: $A = \frac{d_1 d_2}{2}$ and the perimeter of a rhombus is: $P = 2\sqrt{(d_1)^2 + (d_2)^2}$.

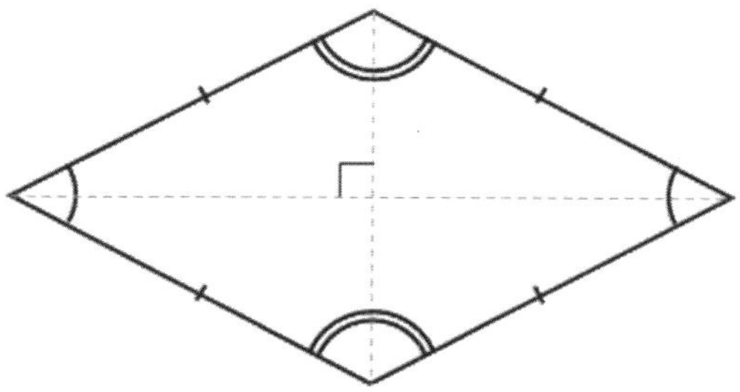

Square

A **square** is a quadrilateral with four right angles and four congruent sides. Squares satisfy the criteria of all other types of quadrilaterals. The diagonals of a square are congruent and perpendicular to each other. Squares have four lines of symmetry (through each pair of opposing midpoints and along each of the diagonals) as well as 90° rotational symmetry about the midpoint.

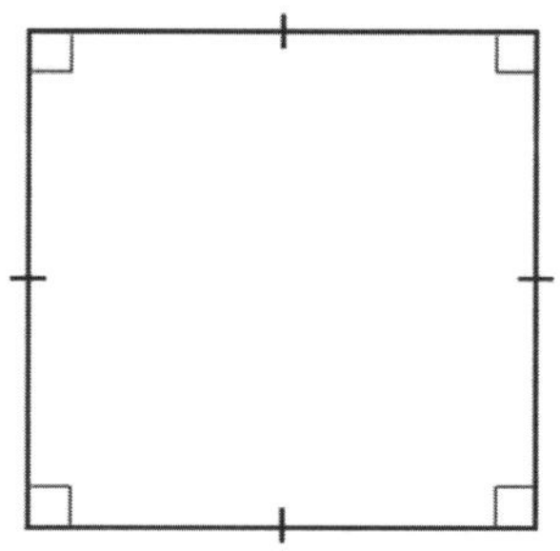

The **area of a square** is found by using the formula $A = s^2$, where s is the length of one side. The **perimeter of a square** is found by using the formula $P = 4s$, where s is the length of one side. Because all four sides are equal in a square, it is faster to multiply the length of one side by 4 than to add the same number four times. You could use the formulas for rectangles and get the same answer.

Review Video: Area and Perimeter of Rectangles and Squares
Visit mometrix.com/academy and enter code: 428109

Hierarchy of Quadrilaterals

The hierarchy of quadrilaterals is as follows:

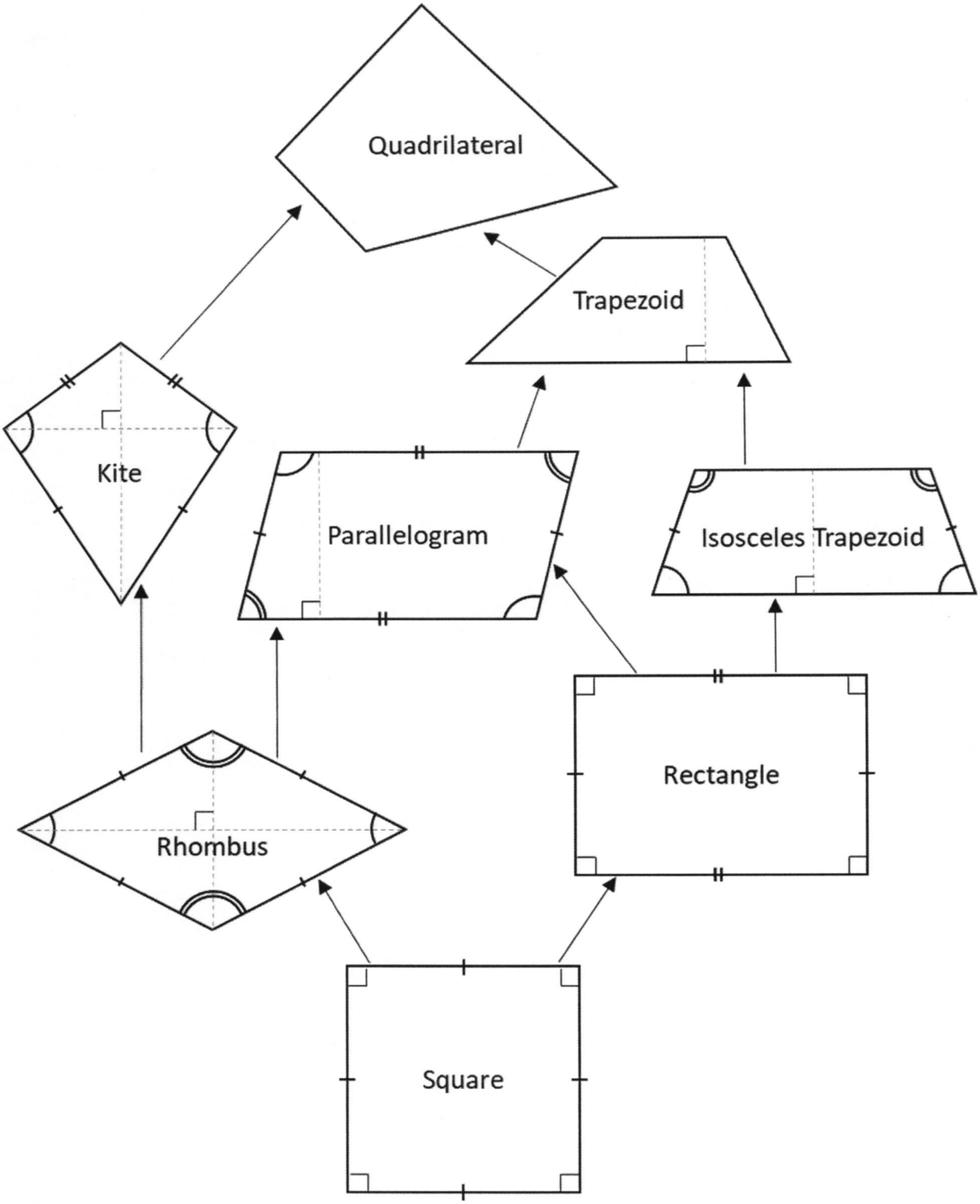

Circles

CIRCLES

The **center** of a circle is the single point from which every point on the circle is **equidistant**. The **radius** is a line segment that joins the center of the circle and any one point on the circle. All radii of a circle are equal. Circles that have the same center but not the same length of radii are **concentric**. The **diameter** is a line segment that passes through the center of the circle and has both endpoints on the circle. The length of the diameter is exactly twice the length of the radius. Point O in the diagram below is the center of the circle, segments $\overline{OX}$, $\overline{OY}$, and $\overline{OZ}$ are radii; and segment $\overline{XZ}$ is a diameter.

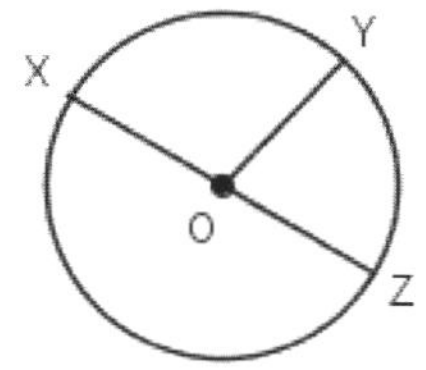

Review Video: Points of a Circle
Visit mometrix.com/academy and enter code: 420746

Review Video: Diameter, Radius, and Circumference
Visit mometrix.com/academy and enter code: 448988

The **area of a circle** is found by the formula $A = \pi r^2$, where r is the length of the radius. If the diameter of the circle is given, remember to divide it in half to get the length of the radius before proceeding.

The **circumference** of a circle is found by the formula $C = 2\pi r$, where r is the radius. Again, remember to convert the diameter if you are given that measure rather than the radius.

Review Video: Area and Circumference of a Circle
Visit mometrix.com/academy and enter code: 243015

INSCRIBED AND CIRCUMSCRIBED FIGURES

These terms can both be used to describe a given arrangement of figures, depending on perspective. If each of the vertices of figure A lie on figure B, then it can be said that figure A is **inscribed** in figure B, but it can also be said that figure B is **circumscribed** about figure A. The following table and examples help to illustrate the concept. Note that the figures cannot both be circles, as they would be completely overlapping and neither would be inscribed or circumscribed.

Given	Description	Equivalent Description	Figures
Each of the sides of a pentagon is tangent to a circle	The circle is inscribed in the pentagon	The pentagon is circumscribed about the circle	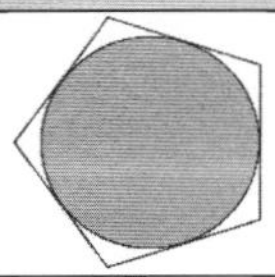
Each of the vertices of a pentagon lie on a circle	The pentagon is inscribed in the circle	The circle is circumscribed about the pentagon	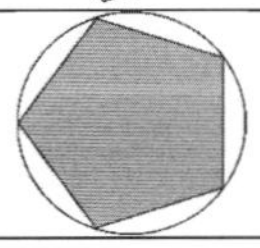

Circle Properties

Arcs

An **arc** is a portion of a circle. Specifically, an arc is the set of points between and including two points on a circle. An arc does not contain any points inside the circle. When a segment is drawn from the endpoints of an arc to the center of the circle, a sector is formed. A **minor arc** is an arc that has a measure less than 180°. A **major arc** is an arc that has a measure of at least 180°. Every minor arc has a corresponding major arc that can be found by subtracting the measure of the minor arc from 360°. A **semicircle** is an arc whose endpoints are the endpoints of the diameter of a circle. A semicircle is exactly half of a circle.

Arc length is the length of that portion of the circumference between two points on the circle. The formula for arc length is $s = \frac{\pi r\theta}{180°}$, where s is the arc length, r is the length of the radius, and θ is the angular measure of the arc in degrees, or $s = r\theta$, where θ is the angular measure of the arc in radians (2π radians $= 360$ degrees).

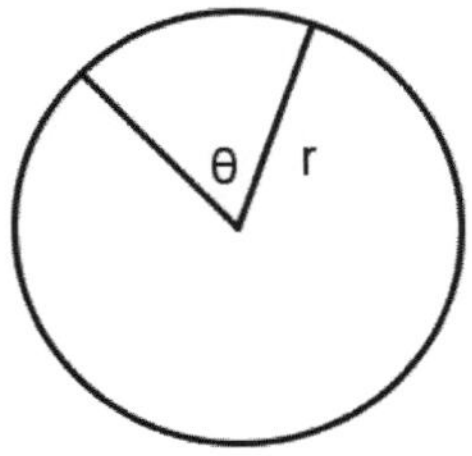

Angles of Circles

A **central angle** is an angle whose vertex is the center of a circle and whose legs intercept an arc of the circle. The measure of a central angle is equal to the measure of the minor arc it intercepts.

An **inscribed angle** is an angle whose vertex lies on a circle and whose legs contain chords of that circle. The portion of the circle intercepted by the legs of the angle is called the intercepted arc. The measure of the intercepted arc is exactly twice the measure of the inscribed angle. In the following diagram, angle ABC is an inscribed angle. $\widehat{AC} = 2(\text{m}\angle ABC)$.

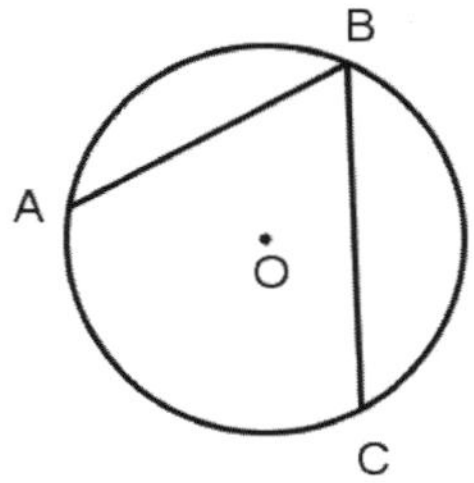

Any angle inscribed in a semicircle is a right angle. The intercepted arc is 180°, making the inscribed angle half that, or 90°. In the diagram below, angle ABC is inscribed in semicircle ABC, making angle ABC equal to 90°.

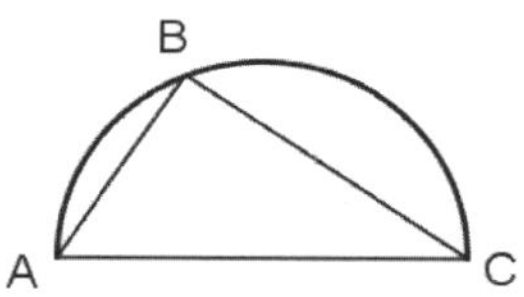

Review Video: Arcs and Angles of Circles
Visit mometrix.com/academy and enter code: 652838

SECANTS, CHORDS, AND TANGENTS

A **secant** is a line that intersects a circle in two points. The segment of a secant line that is contained within the circle is called a **chord**. Two secants may intersect inside the circle, on the circle, or outside the circle. When the two secants intersect on the circle, an inscribed angle is formed. When two secants intersect inside a circle, the measure of each of two vertical angles is equal to half the sum of the two intercepted arcs. Consider the following diagram where $m\angle AEB = \frac{1}{2}(\widehat{AB} + \widehat{CD})$ and $m\angle BEC = \frac{1}{2}(\widehat{BC} + \widehat{AD})$.

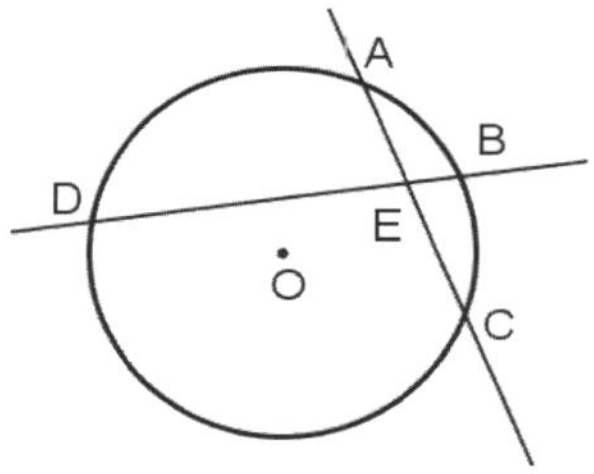

When two secants intersect outside a circle, the measure of the angle formed is equal to half the difference of the two arcs that lie between the two secants. In the diagram below, $m\angle AEB = \frac{1}{2}(\widehat{AB} - \widehat{CD})$.

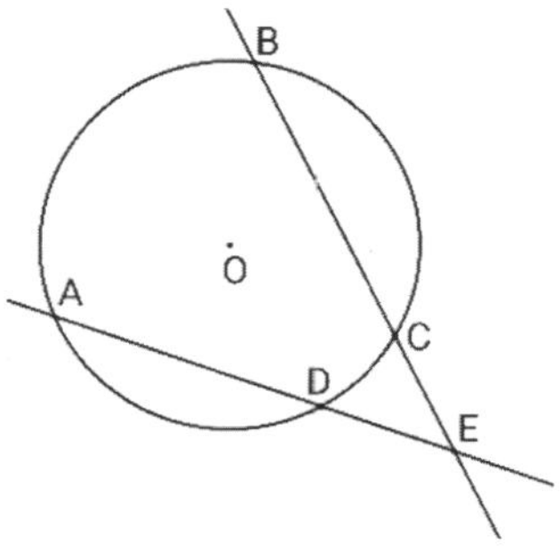

A **tangent** is a line in the same plane as a circle that touches the circle in exactly one point. The point at which a tangent touches a circle is called the **point of tangency**. While a line segment can be tangent to a circle as part of a line that is tangent, it is improper to say a tangent can be simply a line segment that touches the circle in exactly one point.

In the diagram below, $\overleftrightarrow{EB}$ is a secant and contains chord $\overline{EB}$, and $\overleftrightarrow{CD}$ is tangent to circle A. Notice that $\overline{FB}$ is not tangent to the circle. $\overline{FB}$ is a line segment that touches the circle in exactly one point, but if the segment were extended, it would touch the circle in a second point. In the diagram below, point B is the point of tangency.

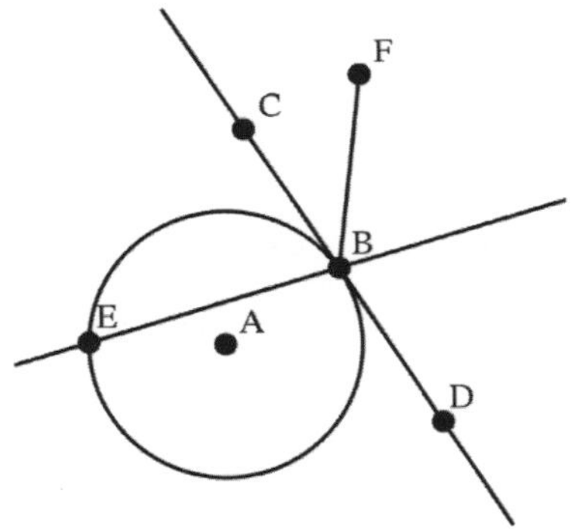

Review Video: Secants, Chords, and Tangents
Visit mometrix.com/academy and enter code: 258360

Review Video: Tangent Lines of a Circle
Visit mometrix.com/academy and enter code: 780167

Sectors

A **sector** is the portion of a circle formed by two radii and their intercepted arc. While the arc length is exclusively the points that are also on the circumference of the circle, the sector is the entire area bounded by the arc and the two radii.

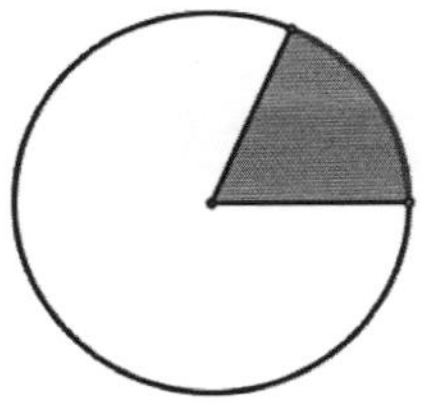

The **area of a sector** of a circle is found by the formula, $A = \frac{\theta r^2}{2}$, where A is the area, θ is the measure of the central angle in radians, and r is the radius. To find the area with the central angle in degrees, use the formula, $A = \frac{\theta \pi r^2}{360}$, where θ is the measure of the central angle and r is the radius.

Transformations

Transformations

Rotation

A **rotation** is a transformation that turns a figure around a point called the **center of rotation**, which can lie anywhere in the plane. If a line is drawn from a point on a figure to the center of rotation, and another line is drawn from the center to the rotated image of that point, the angle between the two lines is the **angle of rotation**. The vertex of the angle of rotation is the center of rotation.

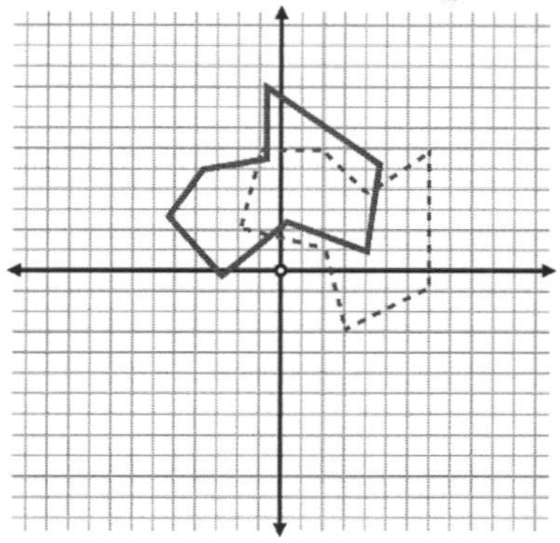

Review Video: Rotation
Visit mometrix.com/academy and enter code: 602600

Translation and Dilation

A **translation** is a transformation which slides a figure from one position in the plane to another position in the plane. The original figure and the translated figure have the same size, shape, and orientation. A **dilation** is a transformation which proportionally stretches or shrinks a figure by a **scale factor**. The dilated image is the same shape and orientation as the original image but a different size. A polygon and its dilated image are similar.

Translation

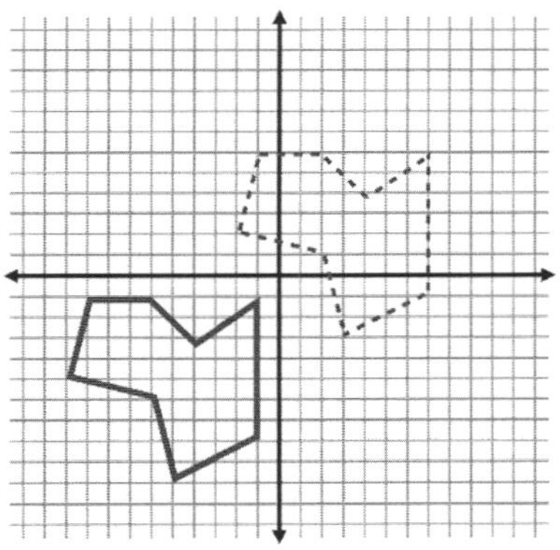

Dilation

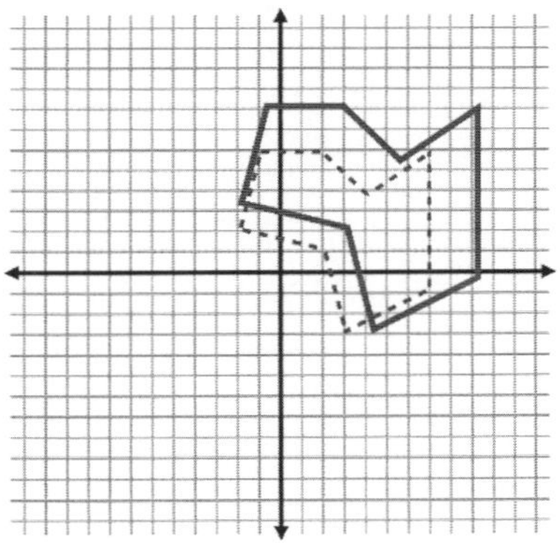

Review Video: Translation
Visit mometrix.com/academy and enter code: 718628

Review Video: Dilation
Visit mometrix.com/academy and enter code: 471630

A **reflection of a figure over a line** (a "flip") creates a congruent image that is the same distance from the line as the original figure but on the opposite side. The **line of reflection** is the perpendicular bisector of any line segment drawn from a point on the original figure to its reflected image (unless the point and its reflected image happen to be the same point, which happens when a figure is reflected over one of its own sides). A **reflection of a figure over a point** (an inversion) in two dimensions is the same as the rotation of the figure 180° about that point. The image of the figure is congruent to the original figure. The **point of reflection** is the midpoint of a line segment which connects a point in the figure to its image (unless the point and its reflected image happen to be the same point, which happens when a figure is reflected in one of its own points).

Reflection of a figure over a line

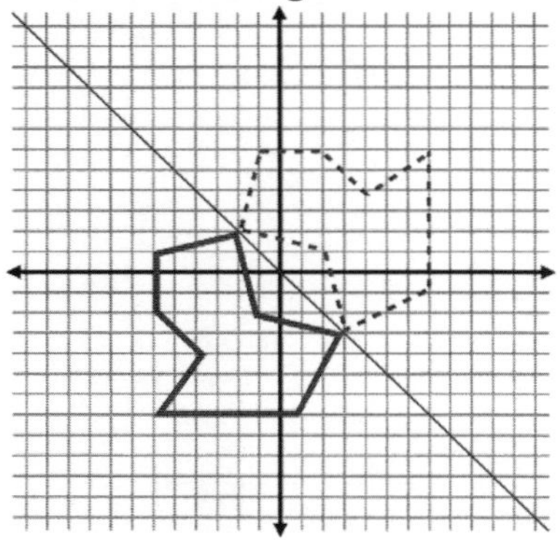

Reflection of a figure over a point

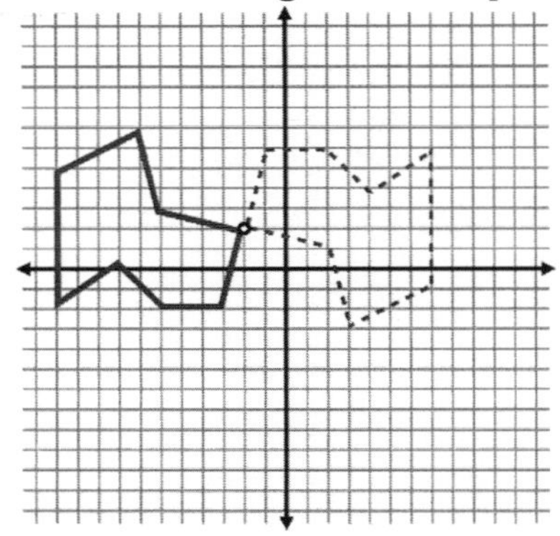

Review Video: Reflection
Visit mometrix.com/academy and enter code: 955068

3D Shapes

3D SHAPES

SOLIDS

The **surface area of a solid object** is the area of all sides or exterior surfaces. For objects such as prisms and pyramids, a further distinction is made between base surface area (B) and lateral surface area (LA). For a prism, the total surface area (SA) is $SA = LA + 2B$. For a pyramid or cone, the total surface area is $SA = LA + B$.

The **surface area of a sphere** can be found by the formula $A = 4\pi r^2$, where r is the radius. The volume is given by the formula $V = \frac{4}{3}\pi r^3$, where r is the radius. Both quantities are generally given in terms of π.

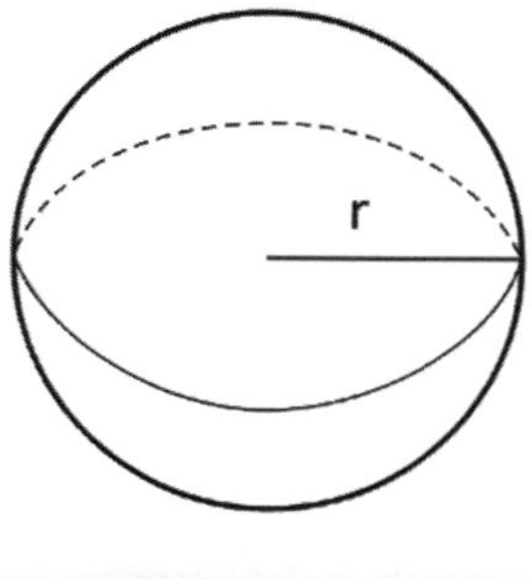

Review Video: Volume and Surface Area of a Sphere
Visit mometrix.com/academy and enter code: 786928

Review Video: How to Calculate the Volume of 3D Objects
Visit mometrix.com/academy and enter code: 163343

The **volume of any prism** is found by the formula $V = Bh$, where B is the area of the base, and h is the height (perpendicular distance between the bases). The surface area of any prism is the sum of the areas of both bases and all sides. It can be calculated as $SA = 2B + Ph$, where P is the perimeter of the base.

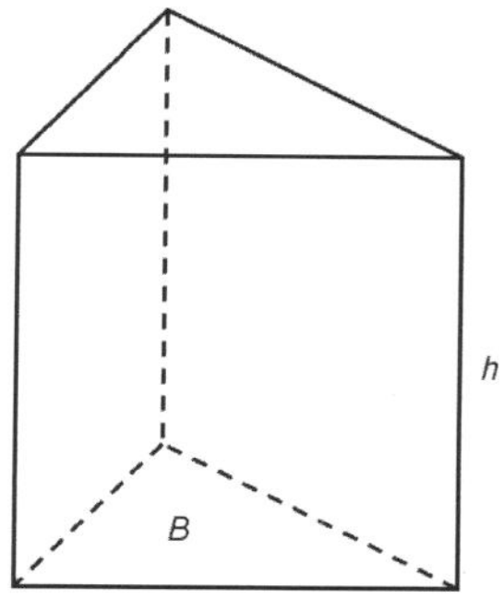

Review Video: Volume and Surface Area of a Prism
Visit mometrix.com/academy and enter code: 420158

For a **rectangular prism**, the volume can be found by the formula $V = lwh$, where V is the volume, l is the length, w is the width, and h is the height. The surface area can be calculated as $SA = 2lw + 2hl + 2wh$ or $SA = 2(lw + hl + wh)$.

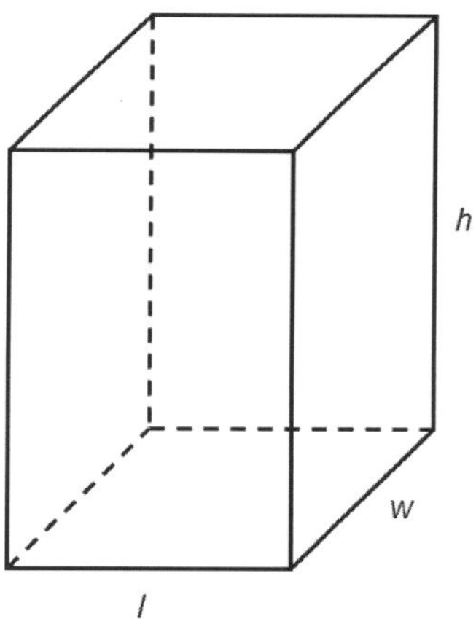

Review Video: Volume and Surface Area of a Rectangular Prism
Visit mometrix.com/academy and enter code: 282814

The **volume of a cube** can be found by the formula $V = s^3$, where s is the length of a side. The surface area of a cube is calculated as $SA = 6s^2$, where SA is the total surface area and s is the length of a side. These formulas are the same as the ones used for the volume and surface area of a rectangular prism, but simplified since all three quantities (length, width, and height) are the same.

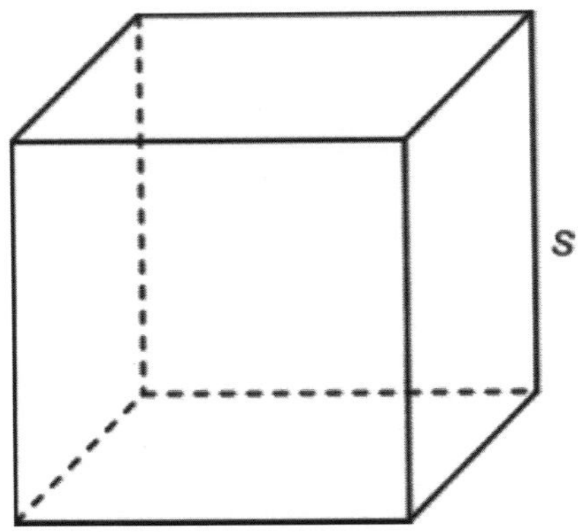

Review Video: Volume and Surface Area of a Cube
Visit mometrix.com/academy and enter code: 664455

The **volume of a cylinder** can be calculated by the formula $V = \pi r^2 h$, where r is the radius, and h is the height. The surface area of a cylinder can be found by the formula $SA = 2\pi r^2 + 2\pi rh$. The first term is the base area multiplied by two, and the second term is the perimeter of the base multiplied by the height.

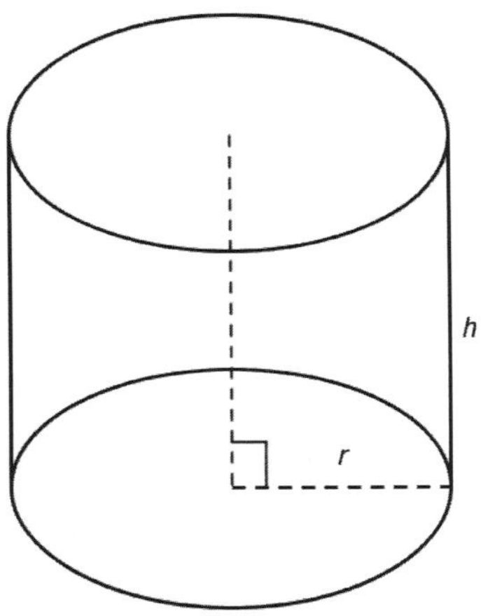

Review Video: Volume and Surface Area of a Right Circular Cylinder
Visit mometrix.com/academy and enter code: 226463

The **volume of a pyramid** is found by the formula $V = \frac{1}{3}Bh$, where B is the area of the base, and h is the height (perpendicular distance from the vertex to the base). Notice this formula is the same as $\frac{1}{3}$ times the volume of a prism. Like a prism, the base of a pyramid can be any shape.

Finding the **surface area of a pyramid** is not as simple as the other shapes we've looked at thus far. If the pyramid is a right pyramid, meaning the base is a regular polygon and the vertex is directly over the center of that polygon, the surface area can be calculated as $SA = B + \frac{1}{2}Ph_s$, where P is the perimeter of the base, and h_s is the slant height (distance from the vertex to the midpoint of one side of the base). If the pyramid is irregular, the area of each triangle side must be calculated individually and then summed, along with the base.

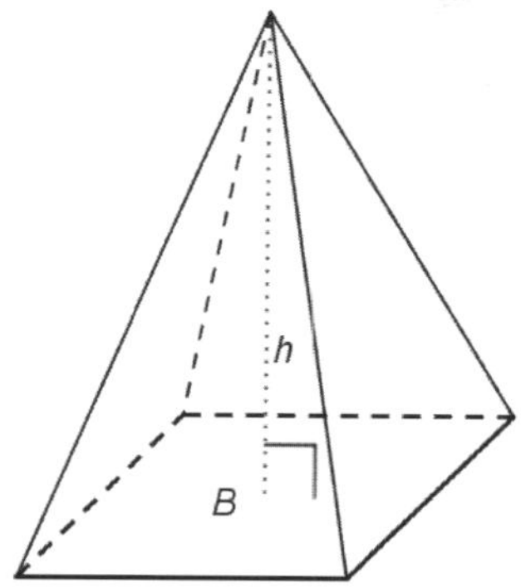

Review Video: Volume and Surface Area of a Pyramid
Visit mometrix.com/academy and enter code: 621932

The **volume of a cone** is found by the formula $V = \frac{1}{3}\pi r^2 h$, where r is the radius, and h is the height. Notice this is the same as $\frac{1}{3}$ times the volume of a cylinder. The surface area can be calculated as $SA = \pi r^2 + \pi r s$, where s is the slant height. The slant height can be calculated using the Pythagorean theorem to be $\sqrt{r^2 + h^2}$, so the surface area formula can also be written as $SA = \pi r^2 + \pi r\sqrt{r^2 + h^2}$.

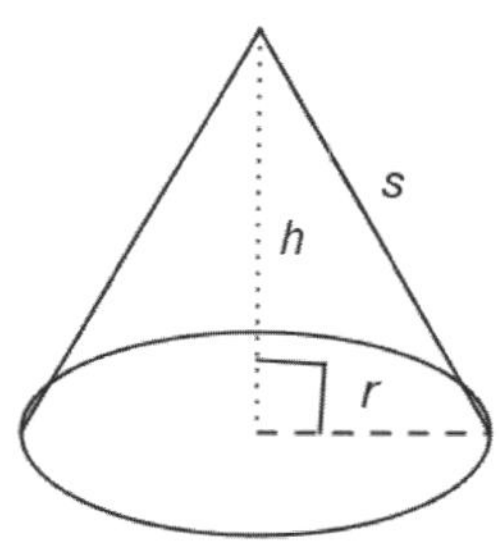

Review Video: Volume and Surface Area of a Right Circular Cone
Visit mometrix.com/academy and enter code: 573574

Conic Sections

Conic sections are a family of shapes that can be thought of as cross sections of a pair of infinite right cones stacked vertex to vertex. This is easiest to see with a visual representation:

A three-dimensional look at representative conic sections. (Note that a hyperbola intersects both cones.)

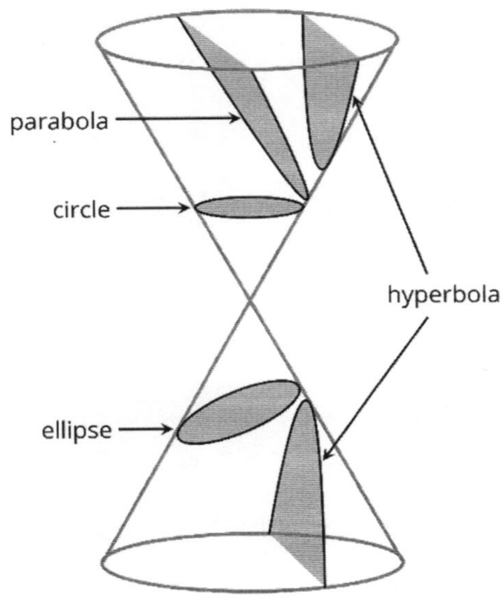

A side-on look at representative conic sections. (Note that the parabola is parallel to the slant of the cones.)

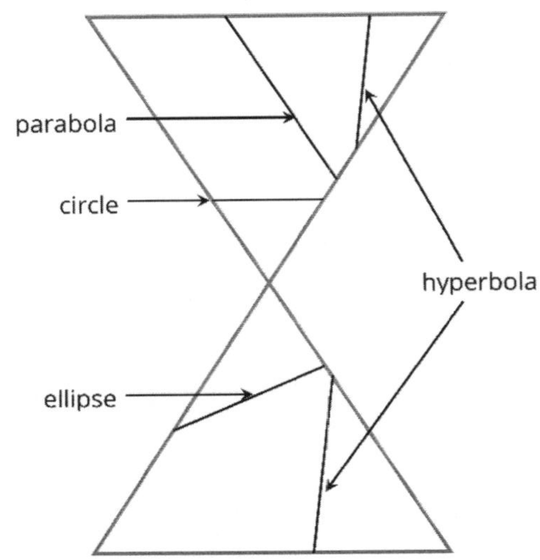

In short, a circle is a horizontal cross section, a parabola is a cross section parallel to the slant of the cone, an ellipse is a cross section at an angle *less than* the slant of the cone, and a hyperbola is a cross section at an angle *greater than* the slant of the cone.

Parabolas, Ellipses, and Hyperbolas

Parabolas

A **parabola** is the set of all points in a plane that are equidistant from a fixed line, called the **directrix**, and a fixed point not on the line, called the **focus**. The **axis** is the line perpendicular to the directrix that passes through the focus.

For parabolas that open up or down, the standard equation is $(x-h)^2 = 4c(y-k)$, where h, c, and k are coefficients. If c is positive, the parabola opens up. If c is negative, the parabola opens down. The vertex is the point (h,k). The directrix is the line having the equation $y = -c + k$, and the focus is the point $(h, c+k)$.

For parabolas that open left or right, the standard equation is $(y-k)^2 = 4c(x-h)$, where k, c, and h are coefficients. If c is positive, the parabola opens to the right. If c is negative, the parabola opens to the left. The vertex is the point (h, k). The directrix is the line having the equation $x = -c + h$, and the focus is the point $(c + h, k)$.

> **Review Video: Parabolas**
> Visit mometrix.com/academy and enter code: 129187
>
> **Review Video: Vertex of a Parabola**
> Visit mometrix.com/academy and enter code: 272300
>
> **Review Video: How to Convert Between Standard Form and Vertex Form**
> Visit mometrix.com/academy and enter code: 916601

ELLIPSES AND HYPERBOLAS

ELLIPSE

An **ellipse** is the set of all points in a plane whose total distance from two fixed points called the **foci** (singular: focus) is constant, and whose center is the midpoint between the foci.

The standard equation of an ellipse that is wider than it is tall is $\frac{(x-h)^2}{a^2} + \frac{(y-k)^2}{b^2} = 1$, where a and b are coefficients. The center is the point (h, k) and the foci are the points $(h + c, k)$ and $(h - c, k)$, where $c^2 = a^2 - b^2$ and $a^2 > b^2$.

The major axis has length $2a$, and the minor axis has length $2b$.

Eccentricity (e) is a measure of how elongated an ellipse is, and is the ratio of the distance between the foci to the length of the major axis. Eccentricity will have a value between 0 and 1. The closer to 0 the eccentricity is, the closer the ellipse is to being a circle. The formula for eccentricity is $e = \frac{c}{a}$.

> **Review Video: Ellipse**
> Visit mometrix.com/academy and enter code: 703899

HYPERBOLA

A **hyperbola** is the set of all points in a plane, whose distance from two fixed points, called foci, has a constant difference.

The standard equation of a horizontal hyperbola is $\frac{(x-h)^2}{a^2} - \frac{(y-k)^2}{b^2} = 1$, where a, b, h, and k are real numbers. The center is the point (h, k), the vertices are the points $(h + a, k)$ and $(h - a, k)$, and the foci are the points that every point on one of the parabolic curves is equidistant from. The foci are found using the formulas $(h + c, k)$ and $(h - c, k)$, where $c^2 = a^2 + b^2$. The asymptotes are two lines the graph of the hyperbola approaches but never reaches, and are given by the equations $y = \left(\frac{b}{a}\right)(x - h) + k$ and $y = -\left(\frac{b}{a}\right)(x - h) + k$.

The standard equation of a vertical hyperbola is $\frac{(y-k)^2}{a^2} - \frac{(x-h)^2}{b^2} = 1$, where a, b, k, and h are real numbers. The center is the point (h, k), the vertices are the points $(h, k + a)$ and $(h, k - a)$, and the foci are the points that every point on one of the hyperbolic curves is equidistant from and are found using the formulas $(h, k + c)$ and $(h, k - c)$, where $c^2 = a^2 + b^2$. The asymptotes are two lines the graph of the hyperbola approaches but never reaches, and are given by the equations $y = \left(\frac{a}{b}\right)(x-h)+k$ and $y = -\left(\frac{a}{b}\right)(x-h)+k$.

Review Video: Hyperbolas
Visit mometrix.com/academy and enter code: 509429

Trigonometric Formulas

Pythagorean Theorem

The side of a triangle opposite the right angle is called the **hypotenuse**. The other two sides are called the legs. The Pythagorean theorem states a relationship among the legs and hypotenuse of a right triangle: $(a^2 + b^2 = c^2)$, where a and b are the lengths of the legs of a right triangle, and c is the length of the hypotenuse. Note that this formula will only work with right triangles.

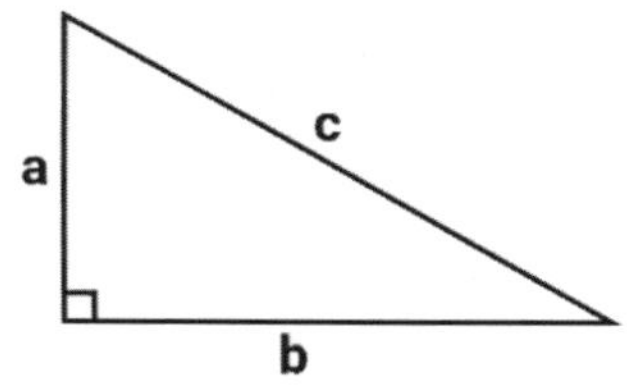

Review Video: Pythagorean Theorem
Visit mometrix.com/academy and enter code: 906576

Trigonometric Formulas

In the diagram below, angle C is the right angle, and side c is the hypotenuse. Side a is the side opposite to angle A and side b is the side opposite to angle B. Using ratios of side lengths as a means to calculate the sine, cosine, and tangent of an acute angle only works for right triangles.

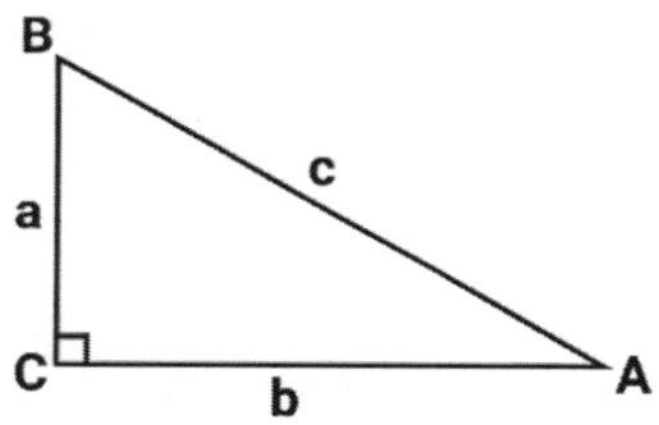

$$\sin A = \frac{\text{opposite side}}{\text{hypotenuse}} = \frac{a}{c} \qquad \csc A = \frac{1}{\sin A} = \frac{\text{hypotenuse}}{\text{opposite side}} = \frac{c}{a}$$

$$\cos A = \frac{\text{adjacent side}}{\text{hypotenuse}} = \frac{b}{c} \qquad \sec A = \frac{1}{\cos A} = \frac{\text{hypotenuse}}{\text{adjacent side}} = \frac{c}{b}$$

$$\tan A = \frac{\text{opposite side}}{\text{adjacent side}} = \frac{a}{b} \qquad \cot A = \frac{1}{\tan A} = \frac{\text{adjacent side}}{\text{opposite side}} = \frac{b}{a}$$

Laws of Sines and Cosines

The **law of sines** states that $\frac{\sin A}{a} = \frac{\sin B}{b} = \frac{\sin C}{c}$, where A, B, and C are the angles of a triangle, and a, b, and c are the sides opposite their respective angles. This formula will work with all triangles, not just right triangles.

The **law of cosines** is given by the formula $c^2 = a^2 + b^2 - 2ab(\cos C)$, where a, b, and c are the sides of a triangle, and C is the angle opposite side c. This is a generalized form of the Pythagorean theorem that can be used on any triangle.

Review Video: Law of Sines
Visit mometrix.com/academy and enter code: 206844

Review Video: Law of Cosines
Visit mometrix.com/academy and enter code: 158911

Chapter Quiz

Ready to see how well you retained what you just read? Scan the QR code to go directly to the chapter quiz interface for this study guide. If you're using a computer, simply visit the online resources page at **mometrix.com/resources719/orelamgmath** and click the Chapter Quizzes link.

Statistics, Probability, and Discrete Mathematics

Displaying Data

DISPLAYING INFORMATION

FREQUENCY TABLES

Frequency tables show how frequently each unique value appears in a set. A **relative frequency table** is one that shows the proportions of each unique value compared to the entire set. Relative frequencies are given as percentages; however, the total percent for a relative frequency table will not necessarily equal 100 percent due to rounding. An example of a frequency table with relative frequencies is below.

Favorite Color	Frequency	Relative Frequency
Blue	4	13%
Red	7	22%
Green	3	9%
Purple	6	19%
Cyan	12	38%

Review Video: Data Interpretation of Graphs
Visit mometrix.com/academy and enter code: 200439

CIRCLE GRAPHS

Circle graphs, also known as *pie charts*, provide a visual depiction of the relationship of each type of data compared to the whole set of data. The circle graph is divided into sections by drawing radii to create central angles whose percentage of the circle is equal to the individual data's percentage of the whole set. Each 1% of data is equal to 3.6° in the circle graph. Therefore, data represented by a 90° section of the circle graph makes up 25% of the whole. When complete, a circle graph often looks like a pie cut into uneven wedges. The pie chart below shows the data from the frequency table referenced earlier where people were asked their favorite color.

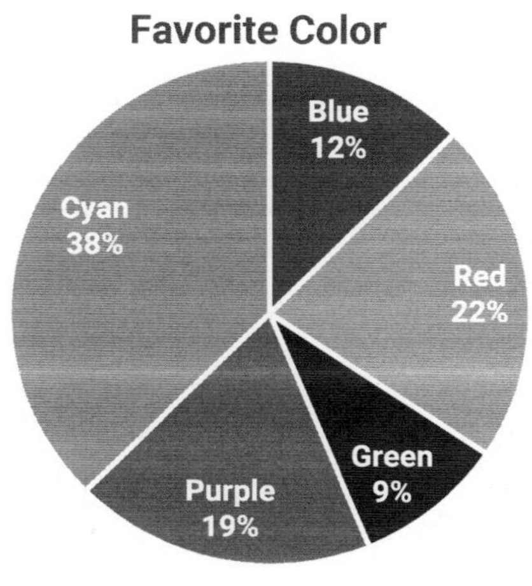

PICTOGRAPHS

A **pictograph** is a graph, generally in the horizontal orientation, that uses pictures or symbols to represent the data. Each pictograph must have a key that defines the picture or symbol and gives the quantity each picture or symbol represents. Pictures or symbols on a pictograph are not always shown as whole elements. In this case, the fraction of the picture or symbol shown represents the same fraction of the quantity a whole picture or symbol stands for. For example, a row with $3\frac{1}{2}$ ears of corn, where each ear of corn represents 100 stalks of corn in a field, would equal $3\frac{1}{2} \times 100 = 350$ stalks of corn in the field.

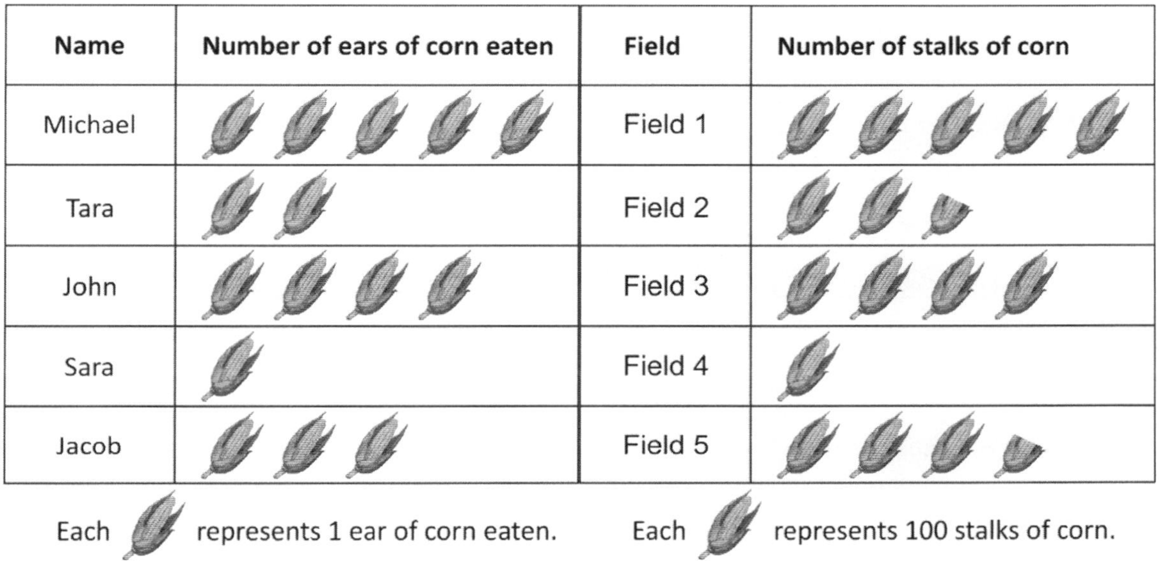

Review Video: Pictographs
Visit mometrix.com/academy and enter code: 147860

LINE GRAPHS

Line graphs have one or more lines of varying styles (solid or broken) to show the different values for a set of data. The individual data are represented as ordered pairs, much like on a Cartesian plane. In this case, the x- and y-axes are defined in terms of their units, such as dollars or time. The individual plotted points are joined by line segments to show whether the value of the data is increasing (line sloping upward), decreasing (line sloping downward), or staying the same (horizontal line). Multiple sets of data can be graphed on the same line graph to give an easy visual comparison. An example of this would be graphing achievement test scores for different groups of students over the same time period to see which group had the greatest increase or decrease in performance from year to year (as shown below).

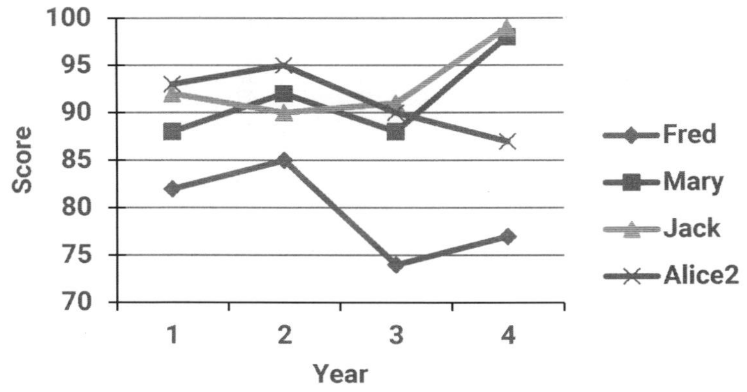

Review Video: How to Create a Line Graph
Visit mometrix.com/academy and enter code: 480147

LINE PLOTS

A **line plot**, also known as a *dot plot*, has plotted points that are not connected by line segments. In this graph, the horizontal axis lists the different possible values for the data, and the vertical axis lists the number of times the individual value occurs. A single dot is graphed for each value to show the number of times it occurs. This graph is more closely related to a bar graph than a line graph. Do not connect the dots in a line plot or it will misrepresent the data.

Review Video: Line Plot
Visit mometrix.com/academy and enter code: 754610

STEM AND LEAF PLOTS

A **stem and leaf plot** is useful for depicting groups of data that fall into a range of values. Each piece of data is separated into two parts: the first, or left, part is called the stem; the second, or right, part is called the leaf. Each stem is listed in a column from smallest to largest. Each leaf that has the common stem is listed in that stem's row from smallest to largest. For example, in a set of two-digit numbers, the digit in the tens place is the stem, and the digit in the ones place is the leaf. With a stem and leaf plot, you can easily see which subset of numbers (10s, 20s, 30s, etc.) is the largest. This information is also readily available by looking at a histogram, but a stem and leaf plot also allows you to look closer and see exactly which values fall in that range. Using a sample set of test scores (82, 88, 92, 93, 85, 90, 92, 95, 74, 88, 90, 91, 78, 87, 98, 99), we can assemble a stem and leaf plot like the one below.

Test Scores

7	4	8							
8	2	5	7	8	8				
9	0	0	1	2	2	3	5	8	9

Review Video: Stem and Leaf Plots
Visit mometrix.com/academy and enter code: 302339

BAR GRAPHS

A **bar graph** is one of the few graphs that can be drawn correctly in two different configurations – both horizontally and vertically. A bar graph is similar to a line plot in the way the data is organized on the graph. Both axes must have their categories defined for the graph to be useful. Rather than placing a single dot to mark the point of the data's value, a bar, or thick line, is drawn from zero to the exact value of the data, whether it is a number, percentage, or other numerical value. Longer bar lengths correspond to greater data values. To read a bar graph, read the labels for the axes to find the units being reported. Then, look where the bars end in relation to the scale given on the corresponding axis and determine the associated value.

The bar chart below represents the responses from our favorite-color survey.

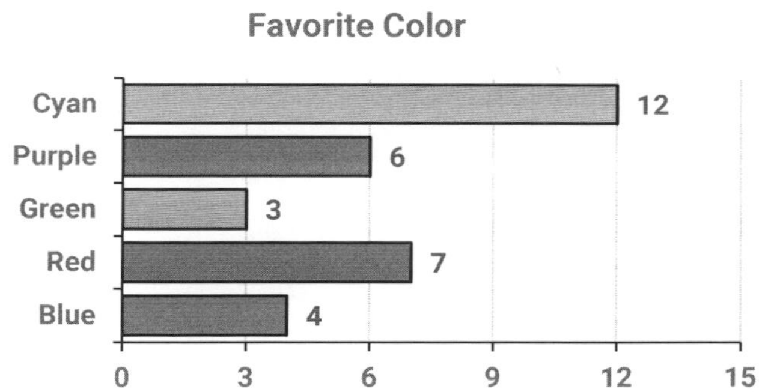

HISTOGRAMS

At first glance, a **histogram** looks like a vertical bar graph. The difference is that a bar graph has a separate bar for each piece of data and a histogram has one continuous bar for each *range* of data. For example, a histogram may have one bar for the range 0–9, one bar for 10–19, etc. While a bar graph has numerical values on one axis, a histogram has numerical values on both axes. Each range is of equal size, and they are ordered left to right from lowest to highest. The height of each column on a histogram represents the number of data values within that range. Like a stem and leaf plot, a histogram makes it easy to glance at the graph and quickly determine which range has the greatest quantity of values. A simple example of a histogram is below.

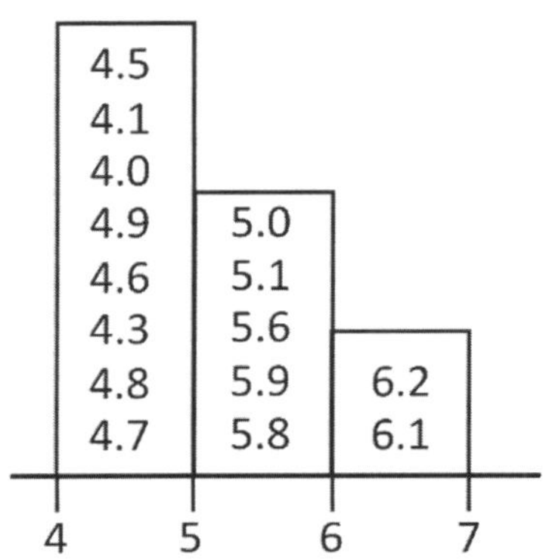

5-NUMBER SUMMARY

The **5-number summary** of a set of data gives a very informative picture of the set. The five numbers in the summary include the minimum value, maximum value, and the three quartiles. This information gives the reader the range and median of the set, as well as an indication of how the data is spread about the median.

BOX AND WHISKER PLOTS

A **box-and-whiskers plot** is a graphical representation of the 5-number summary. To draw a box-and-whiskers plot, plot the points of the 5-number summary on a number line. Draw a box whose ends are through the points for the first and third quartiles. Draw a vertical line in the box through the median to divide the box in half. Draw a line segment from the first quartile point to the minimum value, and from the third quartile point to the maximum value.

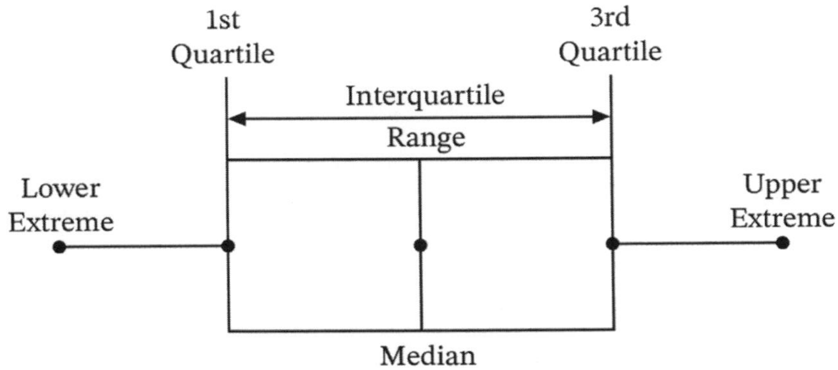

Review Video: Box and Whisker Plots
Visit mometrix.com/academy and enter code: 810817

EXAMPLE

Given the following data (32, 28, 29, 26, 35, 27, 30, 31, 27, 32), we first sort it into numerical order: 26, 27, 27, 28, 29, 30, 31, 32, 32, 35. We can then find the median. Since there are ten values, we take the average of the 5th and 6th values to get 29.5. We find the lower quartile by taking the median of the data smaller than the median. Since there are five values, we take the 3rd value, which is 27. We find the upper quartile by taking the median of the data larger than the overall median, which is 32. Finally, we note our minimum and maximum, which are simply the smallest and largest values in the set: 26 and 35, respectively. Now we can create our box plot:

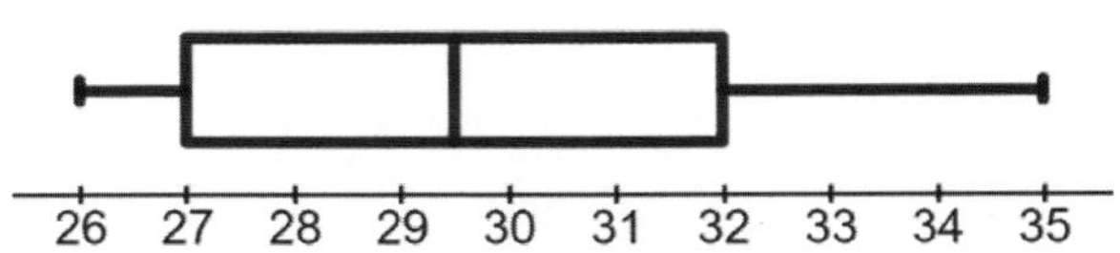

This plot is fairly "long" on the right whisker, showing one or more unusually high values (but not quite outliers). The other quartiles are similar in length, showing a fairly even distribution of data.

INTERQUARTILE RANGE

The **interquartile range, or IQR**, is the difference between the upper and lower quartiles. It measures how the data is dispersed: a high IQR means that the data is more spread out, while a low IQR means that the data is clustered more tightly around the median. To find the IQR, subtract the lower quartile value (Q_1) from the upper quartile value (Q_3).

EXAMPLE

To find the upper and lower quartiles, we first find the median and then take the median of all values above it and all values below it. In the following data set (16, 18, 13, 24, 16, 51, 32, 21, 27, 39), we first rearrange the values in numerical order: 13, 16, 16, 18, 21, 24, 27, 32, 39, 51. There are 10 values, so the median is the average of the 5th and 6th: $\frac{21+24}{2} = \frac{45}{2} = 22.5$. We do not actually need this value to find the upper and lower quartiles. We look at the set of numbers below the median: 13, 16, 16, 18, 21. There are five values, so the 3rd is the median (16), or the value of the lower quartile (Q_1). Then we look at the numbers above the median: 24, 27, 32, 39, 51. Again there are five values, so the 3rd is the median (32), or the value of the upper quartile (Q_3). We find the IQR by subtracting Q_1 from Q_3: $32 - 16 = 16$.

68-95-99.7 RULE

The **68–95–99.7 rule** describes how a normal distribution of data should appear when compared to the mean. This is also a description of a normal bell curve. According to this rule, 68 percent of the data values in a normally distributed set should fall within one standard deviation of the mean (34 percent above and 34 percent below the mean), 95 percent of the data values should fall within two standard deviations of the mean (47.5 percent above and 47.5 percent below the mean), and 99.7 percent of the data values should fall within three standard deviations of the mean, again, equally distributed on either side of the mean. This means that only 0.3 percent of all data values should fall more than three standard deviations from the mean. On the graph below, the normal curve is centered on the y-axis. The x-axis labels are how many standard deviations away from the center you are. Therefore, it is easy to see how the 68-95-99.7 rule can apply.

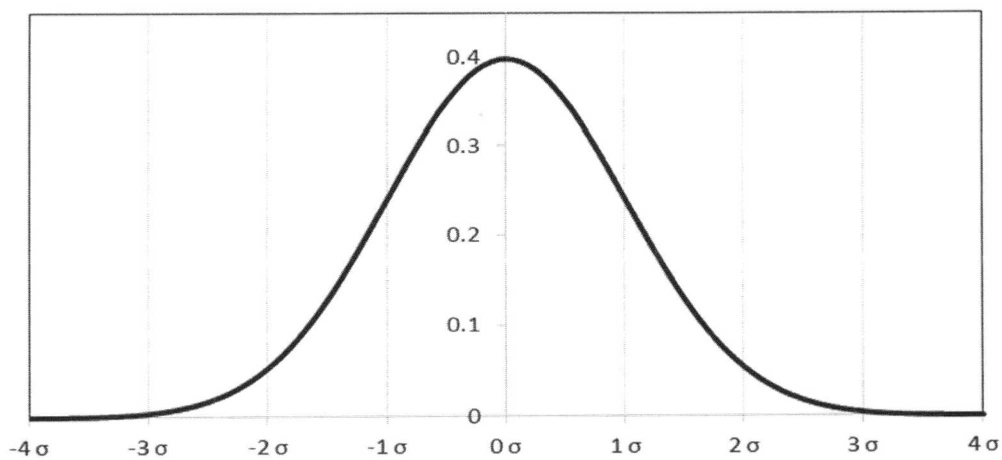

SCATTER PLOTS

BIVARIATE DATA

Bivariate data is simply data from two different variables. (The prefix *bi-* means *two*.) In a *scatter plot*, each value in the set of data is plotted on a grid similar to a Cartesian plane, where each axis represents one of the two variables. By looking at the pattern formed by the points on the grid, you can often determine whether or not there is a relationship between the two variables, and what that relationship is, if it exists. The variables may be directly proportionate, inversely proportionate, or show no proportion at all. It may also be possible to determine if the data is linear, and if so, to find an equation to relate the two variables. The following scatter plot shows the relationship between preference for brand "A" and the age of the consumers surveyed.

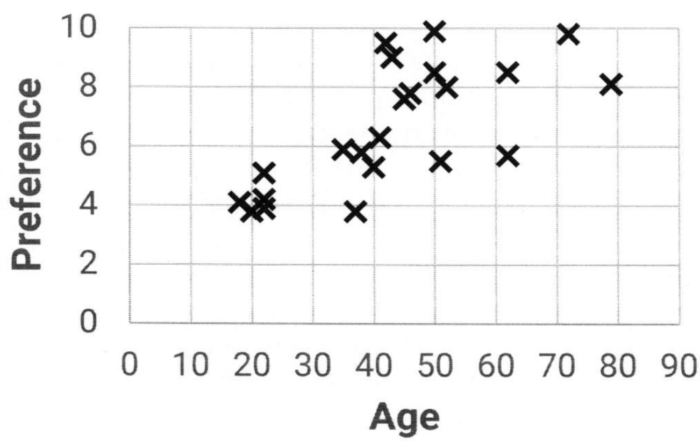

Scatter Plots

Scatter plots are also useful in determining the type of function represented by the data and finding the simple regression. Linear scatter plots may be positive or negative. Nonlinear scatter plots are generally exponential or quadratic. Below are some common types of scatter plots:

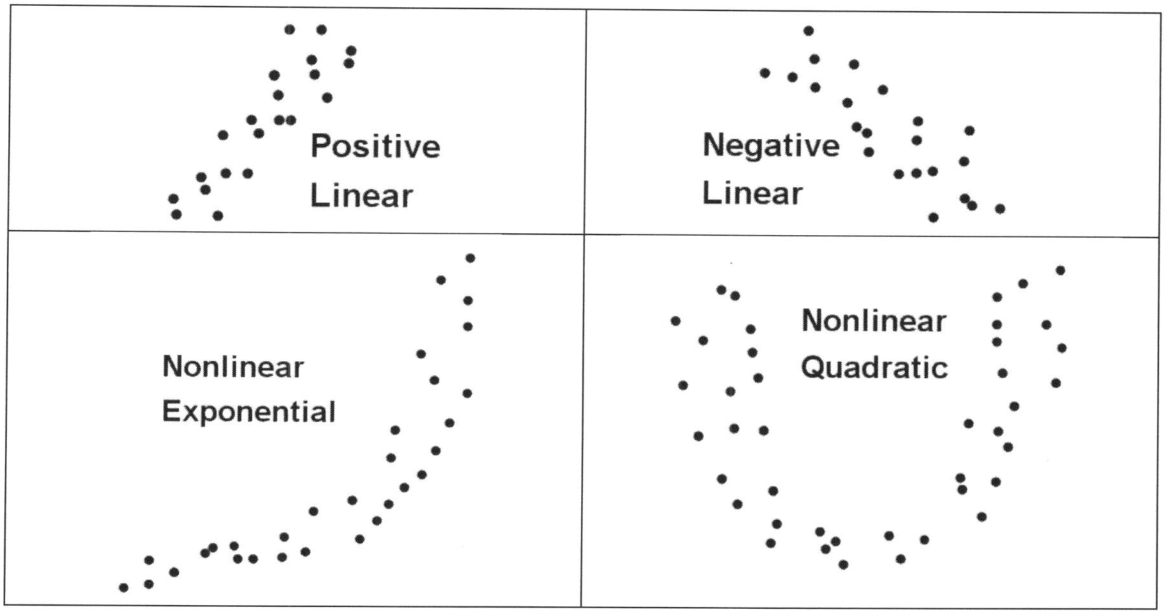

Review Video: Scatter Plot
Visit mometrix.com/academy and enter code: 596526

Statistics

Introduction to Statistics

Statistics is the branch of mathematics that deals with collecting, recording, interpreting, illustrating, and analyzing large amounts of **data**. The following terms are often used in the discussion of data and **statistics**:

- **Data** – the collective name for pieces of information (singular is datum)
- **Quantitative data** – measurements (such as length, mass, and speed) that provide information about quantities in numbers
- **Qualitative data** – information (such as colors, scents, tastes, and shapes) that cannot be measured using numbers
- **Discrete data** – information that can be expressed only by a specific value, such as whole or half numbers. (e.g., since people can be counted only in whole numbers, a population count would be discrete data.)
- **Continuous data** – information (such as time and temperature) that can be expressed by any value within a given range
- **Primary data** – information that has been collected directly from a survey, investigation, or experiment, such as a questionnaire or the recording of daily temperatures. (Primary data that has not yet been organized or analyzed is called **raw data**.)
- **Secondary data** – information that has been collected, sorted, and processed by the researcher

- **Ordinal data** – information that can be placed in numerical order, such as age or weight
- **Nominal data** – information that *cannot* be placed in numerical order, such as names or places

Data Collection

Population

In statistics, the **population** is the entire collection of people, plants, etc., that data can be collected from. For example, a study to determine how well students in local schools perform on a standardized test would have a population of all the students enrolled in those schools, although a study may include just a small sample of students from each school. A **parameter** is a numerical value that gives information about the population, such as the mean, median, mode, or standard deviation. Remember that the symbol for the mean of a population is μ and the symbol for the standard deviation of a population is σ.

Sample

A **sample** is a portion of the entire population. Whereas a parameter helped describe the population, a **statistic** is a numerical value that gives information about the sample, such as mean, median, mode, or standard deviation. Keep in mind that the symbols for mean and standard deviation are different when they are referring to a sample rather than the entire population. For a sample, the symbol for mean is $\bar{x}$ and the symbol for standard deviation is s. The mean and standard deviation of a sample may or may not be identical to that of the entire population due to a sample only being a subset of the population. However, if the sample is random and large enough, statistically significant values can be attained. Samples are generally used when the population is too large to justify including every element or when acquiring data for the entire population is impossible.

Inferential Statistics

Inferential statistics is the branch of statistics that uses samples to make predictions about an entire population. This type of statistic is often seen in political polls, where a sample of the population is questioned about a particular topic or politician to gain an understanding of the attitudes of the entire population of the country. Often, exit polls are conducted on election days using this method. Inferential statistics can have a large margin of error if you do not have a valid sample.

Sampling Distribution

Statistical values calculated from various samples of the same size make up the **sampling distribution**. For example, if several samples of identical size are randomly selected from a large population and then the mean of each sample is calculated, the distribution of values of the means would be a sampling distribution.

The **sampling distribution of the mean** is the distribution of the sample mean, $\bar{x}$, derived from random samples of a given size. It has three important characteristics. First, the mean of the sampling distribution of the mean is equal to the mean of the population that was sampled. Second, assuming the standard deviation is non-zero, the standard deviation of the sampling distribution of the mean equals the standard deviation of the sampled population divided by the square root of the sample size. This is sometimes called the standard error. Finally, as the sample size gets larger, the sampling distribution of the mean gets closer to a normal distribution via the central limit theorem.

Survey Study

A **survey study** is a method of gathering information from a small group in an attempt to gain enough information to make accurate general assumptions about the population. Once a survey study is completed, the results are then put into a summary report.

Survey studies are generally in the format of surveys, interviews, or questionnaires as part of an effort to find opinions of a particular group or to find facts about a group.

It is important to note that the findings from a survey study are only as accurate as the sample chosen from the population.

Correlational Studies

Correlational studies seek to determine how much one variable is affected by changes in a second variable. For example, correlational studies may look for a relationship between the amount of time a student spends studying for a test and the grade that student earned on the test or between student scores on college admissions tests and student grades in college.

It is important to note that correlational studies cannot show a cause and effect, but rather can show only that two variables are or are not potentially correlated.

Experimental Studies

Experimental studies take correlational studies one step farther, in that they attempt to prove or disprove a cause-and-effect relationship. These studies are performed by conducting a series of experiments to test the hypothesis. For a study to be scientifically accurate, it must have both an experimental group that receives the specified treatment and a control group that does not get the treatment. This is the type of study pharmaceutical companies do as part of drug trials for new medications. Experimental studies are only valid when the proper scientific method has been followed. In other words, the experiment must be well-planned and executed without bias in the testing process, all subjects must be selected at random, and the process of determining which subject is in which of the two groups must also be completely random.

Observational Studies

Observational studies are the opposite of experimental studies. In observational studies, the tester cannot change or in any way control all of the variables in the test. For example, a study to determine which gender does better in math classes in school is strictly observational. You cannot change a person's gender, and you cannot change the subject being studied. The big downfall of observational studies is that you have no way of proving a cause-and-effect relationship because you cannot control outside influences. Events outside of school can influence a student's performance in school, and observational studies cannot take that into consideration.

Random Samples

For most studies, a **random sample** is necessary to produce valid results. Random samples should not have any particular influence to cause sampled subjects to behave one way or another. The goal is for the random sample to be a **representative sample**, or a sample whose characteristics give an accurate picture of the characteristics of the entire population. To accomplish this, you must make sure you have a proper **sample size**, or an appropriate number of elements in the sample.

Biases

In statistical studies, biases must be avoided. **Bias** is an error that causes the study to favor one set of results over another. For example, if a survey to determine how the country views the president's job performance only speaks to registered voters in the president's party, the results will be skewed because a disproportionately large number of responders would tend to show approval, while a disproportionately large number of people in the opposite party would tend to express disapproval. **Extraneous variables** are, as the name implies, outside influences that can affect the outcome of a study. They are not always avoidable but could trigger bias in the result.

DATA ANALYSIS

DISPERSION

A **measure of dispersion** is a single value that helps to "interpret" the measure of central tendency by providing more information about how the data values in the set are distributed about the measure of central tendency. The measure of dispersion helps to eliminate or reduce the disadvantages of using the mean, median, or mode as a single measure of central tendency, and give a more accurate picture of the dataset as a whole. To have a measure of dispersion, you must know or calculate the range, standard deviation, or variance of the data set.

RANGE

The **range** of a set of data is the difference between the greatest and lowest values of the data in the set. To calculate the range, you must first make sure the units for all data values are the same, and then identify the greatest and lowest values. If there are multiple data values that are equal for the highest or lowest, just use one of the values in the formula. Write the answer with the same units as the data values you used to do the calculations.

Review Video: Statistical Range
Visit mometrix.com/academy and enter code: 778541

SAMPLE STANDARD DEVIATION

Standard deviation is a measure of dispersion that compares all the data values in the set to the mean of the set to give a more accurate picture. To find the **standard deviation of a sample**, use the formula

$$s = \sqrt{\frac{\sum_{i=1}^{n}(x_i - \bar{x})^2}{n-1}}$$

Note that s is the standard deviation of a sample, x_i represents the individual values in the data set, $\bar{x}$ is the mean of the data values in the set, and n is the number of data values in the set. The higher the value of the standard deviation is, the greater the variance of the data values from the mean. The units associated with the standard deviation are the same as the units of the data values.

Review Video: Standard Deviation
Visit mometrix.com/academy and enter code: 419469

SAMPLE VARIANCE

The **variance of a sample** is the square of the sample standard deviation (denoted s^2). While the mean of a set of data gives the average of the set and gives information about where a specific data value lies in relation to the average, the variance of the sample gives information about the degree to which the data values are spread out and tells you how close an individual value is to the average compared to the other values. The units associated with variance are the same as the units of the data values squared.

PERCENTILE

Percentiles and quartiles are other methods of describing data within a set. **Percentiles** tell what percentage of the data in the set fall below a specific point. For example, achievement test scores are often given in percentiles. A score at the 80th percentile is one which is equal to or higher than 80 percent of the scores in the set. In other words, 80 percent of the scores were lower than that score.

Quartiles are percentile groups that make up quarter sections of the data set. The first quartile is the 25th percentile. The second quartile is the 50th percentile; this is also the median of the dataset. The third quartile is the 75th percentile.

SKEWNESS

Skewness is a way to describe the symmetry or asymmetry of the distribution of values in a dataset. If the distribution of values is symmetrical, there is no skew. In general the closer the mean of a data set is to the median of the data set, the less skew there is. Generally, if the mean is to the right of the median, the data set is *positively skewed*, or right-skewed, and if the mean is to the left of the median, the data set is *negatively skewed*, or left-skewed. However, this rule of thumb is not infallible. When the data values are graphed on a curve, a set with no skew will be a perfect bell curve.

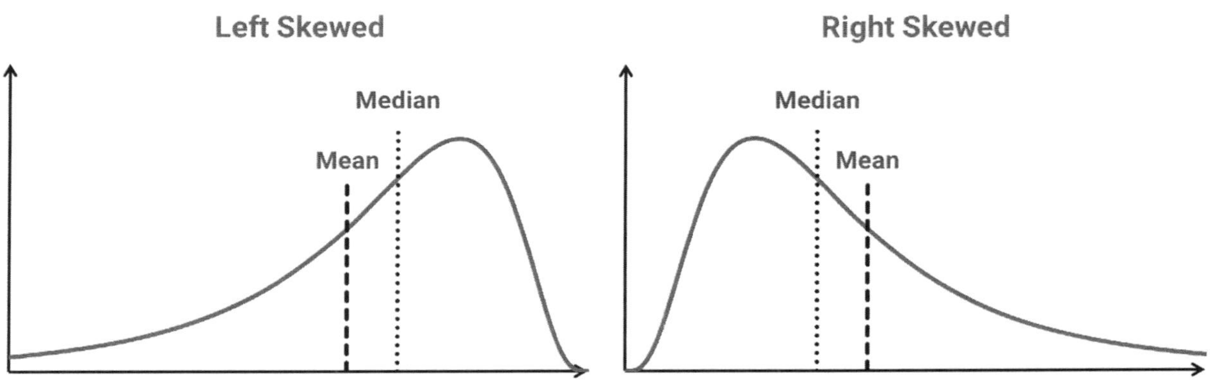

To estimate skew, use the formula:

$$\text{skew} = \frac{\sqrt{n(n-1)}}{n-2}\left(\frac{\frac{1}{n}\sum_{i=1}^{n}(x_i-\bar{x})^3}{\left(\frac{1}{n}\sum_{i=1}^{n}(x_i-\bar{x})^2\right)^{\frac{3}{2}}}\right)$$

Note that n is the datapoints in the set, x_i is the ith value in the set, and $\bar{x}$ is the mean of the set.

Review Video: Skew
Visit mometrix.com/academy and enter code: 661486

UNIMODAL VS. BIMODAL

If a distribution has a single peak, it would be considered **unimodal**. If it has two discernible peaks it would be considered **bimodal**. Bimodal distributions may be an indication that the set of data being considered is actually the combination of two sets of data with significant differences. A

uniform distribution is a distribution in which there is *no distinct peak or variation* in the data. No values or ranges are particularly more common than any other values or ranges.

Outlier

An outlier is an extremely high or extremely low value in the data set. It may be the result of measurement error, in which case, the outlier is not a valid member of the data set. However, it may also be a valid member of the distribution. Unless a measurement error is identified, the experimenter cannot know for certain if an outlier is or is not a member of the distribution. There are arbitrary methods that can be employed to designate an extreme value as an outlier. One method designates an outlier (or possible outlier) to be any value less than $Q_1 - 1.5(IQR)$ or any value greater than $Q_3 + 1.5(IQR)$.

Data Analysis

Simple Regression

In statistics, **simple regression** is using an equation to represent a relation between independent and dependent variables. The independent variable is also referred to as the explanatory variable or the predictor and is generally represented by the variable x in the equation. The dependent variable, usually represented by the variable y, is also referred to as the response variable. The equation may be any type of function – linear, quadratic, exponential, etc. The best way to handle this task is to use the regression feature of your graphing calculator. This will easily give you the curve of best fit and provide you with the coefficients and other information you need to derive an equation.

Line of Best Fit

In a scatter plot, the **line of best fit** is the line that best shows the trends of the data. The line of best fit is given by the equation $\hat{y} = ax + b$, where a and b are the regression coefficients. The regression coefficient a is also the slope of the line of best fit, and b is also the y-coordinate of the point at which the line of best fit crosses the y-axis. Not every point on the scatter plot will be on the line of best fit. The differences between the y-values of the points in the scatter plot and the corresponding y-values according to the equation of the line of best fit are the residuals. The line of best fit is also called the least-squares regression line because it is also the line that has the lowest sum of the squares of the residuals.

Correlation Coefficient

The **correlation coefficient** is the numerical value that indicates how strong the relationship is between the two variables of a linear regression equation. A correlation coefficient of –1 is a perfect negative correlation. A correlation coefficient of +1 is a perfect positive correlation. Correlation coefficients close to –1 or +1 are very strong correlations. A correlation coefficient equal to zero indicates there is no correlation between the two variables. This test is a good indicator of whether or not the equation for the line of best fit is accurate. The formula for the correlation coefficient is

$$r = \frac{\sum_{i=1}^{n}(x_i - \bar{x})(y_i - \bar{y})}{\sqrt{\sum_{i=1}^{n}(x_i - \bar{x})^2}\sqrt{\sum_{i=1}^{n}(y_i - \bar{y})^2}}$$

where r is the correlation coefficient, n is the number of data values in the set, (x_i, y_i) is a point in the set, and $\bar{x}$ and $\bar{y}$ are the means.

Z-Score

A **z-score** is an indication of how many standard deviations a given value falls from the sample mean. To calculate a z-score, use the formula:

$$\frac{x - \bar{x}}{\sigma}$$

In this formula x is the data value, $\bar{x}$ is the mean of the sample data, and σ is the standard deviation of the population. If the z-score is positive, the data value lies above the mean. If the z-score is negative, the data value falls below the mean. These scores are useful in interpreting data such as standardized test scores, where every piece of data in the set has been counted, rather than just a small random sample. In cases where standard deviations are calculated from a random sample of the set, the z-scores will not be as accurate.

Central Limit Theorem

According to the **central limit theorem**, regardless of what the original distribution of a sample is, the distribution of the means tends to get closer and closer to a normal distribution as the sample size gets larger and larger (this is necessary because the sample is becoming more all-encompassing of the elements of the population). As the sample size gets larger, the distribution of the sample mean will approach a normal distribution with a mean of the population mean and a variance of the population variance divided by the sample size.

Measures of Central Tendency

A **measure of central tendency** is a statistical value that gives a reasonable estimate for the center of a group of data. There are several different ways of describing the measure of central tendency. Each one has a unique way it is calculated, and each one gives a slightly different perspective on the data set. Whenever you give a measure of central tendency, always make sure the units are the same. If the data has different units, such as hours, minutes, and seconds, convert all the data to the same unit, and use the same unit in the measure of central tendency. If no units are given in the data, do not give units for the measure of central tendency.

Mean

The **statistical mean** of a group of data is the same as the arithmetic average of that group. To find the mean of a set of data, first convert each value to the same units, if necessary. Then find the sum of all the values, and count the total number of data values, making sure you take into consideration each individual value. If a value appears more than once, count it more than once. Divide the sum of the values by the total number of values and apply the units, if any. Note that the mean does not have to be one of the data values in the set, and may not divide evenly.

$$\text{mean} = \frac{\text{sum of the data values}}{\text{quantity of data values}}$$

For instance, the mean of the data set {88, 72, 61, 90, 97, 68, 88, 79, 86, 93, 97, 71, 80, 84, 89} would be the sum of the fifteen numbers divided by 15:

$$\frac{88 + 72 + 61 + 90 + 97 + 68 + 88 + 79 + 86 + 93 + 97 + 71 + 80 + 84 + 89}{15} = \frac{1242}{15}$$
$$= 82.8$$

While the mean is relatively easy to calculate and averages are understood by most people, the mean can be very misleading if it is used as the sole measure of central tendency. If the data set has outliers (data values that are unusually high or unusually low compared to the rest of the data values), the mean can be very distorted, especially if the data set has a small number of values. If unusually high values are countered with unusually low values, the mean is not affected as much. For example, if five of twenty students in a class get a 100 on a test, but the other 15 students have an average of 60 on the same test, the class average would appear as 70. Whenever the mean is skewed by outliers, it is always a good idea to include the median as an alternate measure of central tendency.

A **weighted mean**, or weighted average, is a mean that uses "weighted" values. The formula is weighted mean $= \frac{w_1x_1+w_2x_2+w_3x_3...+w_nx_n}{w_1+w_2+w_3+\cdots+w_n}$. Weighted values, such as $w_1, w_2, w_3, \dots w_n$ are assigned to each member of the set $x_1, x_2, x_3, \dots x_n$. When calculating the weighted mean, make sure a weight value for each member of the set is used.

Review Video: All About Averages
Visit mometrix.com/academy and enter code: 176521

MEDIAN

The **statistical median** is the value in the middle of the set of data. To find the median, list all data values in order from smallest to largest or from largest to smallest. Any value that is repeated in the set must be listed the number of times it appears. If there are an odd number of data values, the median is the value in the middle of the list. If there is an even number of data values, the median is the arithmetic mean of the two middle values.

For example, the median of the data set {88, 72, 61, 90, 97, 68, 88, 79, 86, 93, 97, 71, 80, 84, 88} is 86 since the ordered set is {61, 68, 71, 72, 79, 80, 84, **86**, 88, 88, 88, 90, 93, 97, 97}.

The big disadvantage of using the median as a measure of central tendency is that is relies solely on a value's relative size as compared to the other values in the set. When the individual values in a set of data are evenly dispersed, the median can be an accurate tool. However, if there is a group of rather large values or a group of rather small values that are not offset by a different group of values, the information that can be inferred from the median may not be accurate because the distribution of values is skewed.

MODE

The **statistical mode** is the data value that occurs the greatest number of times in the data set. It is possible to have exactly one mode, more than one mode, or no mode. To find the mode of a set of data, arrange the data like you do to find the median (all values in order, listing all multiples of data values). Count the number of times each value appears in the data set. If all values appear an equal number of times, there is no mode. If one value appears more than any other value, that value is the mode. If two or more values appear the same number of times, but there are other values that appear fewer times and no values that appear more times, all of those values are the modes.

For example, the mode of the data set {**88**, 72, 61, 90, 97, 68, **88**, 79, 86, 93, 97, 71, 80, 84, **88**} is 88.

The main disadvantage of the mode is that the values of the other data in the set have no bearing on the mode. The mode may be the largest value, the smallest value, or a value anywhere in between in the set. The mode only tells which value or values, if any, occurred the greatest number of times. It does not give any suggestions about the remaining values in the set.

Review Video: Mean, Median, and Mode
Visit mometrix.com/academy and enter code: 286207

Probability

PROBABILITY

Probability is the likelihood of a certain outcome occurring for a given event. An **event** is any situation that produces a result. It could be something as simple as flipping a coin or as complex as launching a rocket. Determining the probability of an outcome for an event can be equally simple or complex. As such, there are specific terms used in the study of probability that need to be understood:

- **Compound event**—an event that involves two or more independent events (rolling a pair of dice and taking the sum)
- **Desired outcome** (or success)—an outcome that meets a particular set of criteria (a roll of 1 or 2 if we are looking for numbers less than 3)
- **Independent events**—two or more events whose outcomes do not affect one another (two coins tossed at the same time)
- **Dependent events**—two or more events whose outcomes affect one another (two cards drawn consecutively from the same deck)
- **Certain outcome**—probability of outcome is 100% or 1
- **Impossible outcome**—probability of outcome is 0% or 0
- **Mutually exclusive outcomes**—two or more outcomes whose criteria cannot all be satisfied in a single event (a coin coming up heads and tails on the same toss)
- **Random variable**—refers to all possible outcomes of a single event which may be discrete or continuous.

Review Video: Intro to Probability
Visit mometrix.com/academy and enter code: 212374

SAMPLE SPACE

The total set of all possible results of a test or experiment is called a **sample space**, or sometimes a universal sample space. The sample space, represented by one of the variables S, Ω, or U (for universal sample space) has individual elements called outcomes. Other terms for outcome that may be used interchangeably include elementary outcome, simple event, or sample point. The number of outcomes in a given sample space could be infinite or finite, and some tests may yield multiple unique sample sets. For example, tests conducted by drawing playing cards from a standard deck would have one sample space of the card values, another sample space of the card suits, and a third sample space of suit-denomination combinations. For most tests, the sample spaces considered will be finite.

An **event**, represented by the variable E, is a portion of a sample space. It may be one outcome or a group of outcomes from the same sample space. If an event occurs, then the test or experiment will generate an outcome that satisfies the requirement of that event. For example, given a standard deck of 52 playing cards as the sample space, and defining the event as the collection of face cards, then the event will occur if the card drawn is a J, Q, or K. If any other card is drawn, the event is said to have not occurred.

For every sample space, each possible outcome has a specific likelihood, or probability, that it will occur. The probability measure, also called the **distribution**, is a function that assigns a real number probability, from zero to one, to each outcome. For a probability measure to be accurate, every outcome must have a real number probability measure that is greater than or equal to zero and less than or equal to one. Also, the probability measure of the sample space must equal one, and the probability measure of the union of multiple outcomes must equal the sum of the individual probability measures.

Probabilities of events are expressed as real numbers from zero to one. They give a numerical value to the chance that a particular event will occur. The probability of an event occurring is the sum of the probabilities of the individual elements of that event. For example, in a standard deck of 52 playing cards as the sample space and the collection of face cards as the event, the probability of drawing a specific face card is $\frac{1}{52} = 0.019$, but the probability of drawing any one of the twelve face cards is $12(0.019) = 0.228$. Note that rounding of numbers can generate different results. If you multiplied 12 by the fraction $\frac{1}{52}$ before converting to a decimal, you would get the answer $\frac{12}{52} = 0.231$.

Theoretical and Experimental Probability

Theoretical probability can usually be determined without actually performing the event. The likelihood of an outcome occurring, or the probability of an outcome occurring, is given by the formula:

$$P(A) = \frac{\text{Number of acceptable outcomes}}{\text{Number of possible outcomes}}$$

Note that $P(A)$ is the probability of an outcome A occurring, and each outcome is just as likely to occur as any other outcome. If each outcome has the same probability of occurring as every other possible outcome, the outcomes are said to be equally likely to occur. The total number of acceptable outcomes must be less than or equal to the total number of possible outcomes. If the two are equal, then the outcome is certain to occur and the probability is 1. If the number of acceptable outcomes is zero, then the outcome is impossible and the probability is 0. For example, if there are 20 marbles in a bag and 5 are red, then the theoretical probability of randomly selecting a red marble is 5 out of 20, $\left(\frac{5}{20} = \frac{1}{4}, 0.25, \text{or } 25\%\right)$.

If the theoretical probability is unknown or too complicated to calculate, it can be estimated by an experimental probability. **Experimental probability**, also called empirical probability, is an estimate of the likelihood of a certain outcome based on repeated experiments or collected data. In other words, while theoretical probability is based on what *should* happen, experimental probability is based on what *has* happened. Experimental probability is calculated in the same way as theoretical probability, except that actual outcomes are used instead of possible outcomes. The more experiments performed or datapoints gathered, the better the estimate should be.

Theoretical and experimental probability do not always line up with one another. Theoretical probability says that out of 20 coin-tosses, 10 should be heads. However, if we were actually to toss 20 coins, we might record just 5 heads. This doesn't mean that our theoretical probability is incorrect; it just means that this particular experiment had results that were different from what was predicted. A practical application of empirical probability is the insurance industry. There are no set functions that define lifespan, health, or safety. Insurance companies look at factors from hundreds of thousands of individuals to find patterns that they then use to set the formulas for insurance premiums.

Review Video: Empirical Probability
Visit mometrix.com/academy and enter code: 513468

Objective and Subjective Probability

Objective probability is based on mathematical formulas and documented evidence. Examples of objective probability include raffles or lottery drawings where there is a pre-determined number of possible outcomes and a predetermined number of outcomes that correspond to an event. Other cases of objective probability include probabilities of rolling dice, flipping coins, or drawing cards. Most gambling games are based on objective probability.

In contrast, **subjective probability** is based on personal or professional feelings and judgments. Often, there is a lot of guesswork following extensive research. Areas where subjective probability is applicable include sales trends and business expenses. Attractions set admission prices based on subjective probabilities of attendance based on varying admission rates in an effort to maximize their profit.

Complement of an Event

Sometimes it may be easier to calculate the possibility of something not happening, or the **complement of an event**. Represented by the symbol $\bar{A}$, the complement of A is the probability that event A does not happen. When you know the probability of event A occurring, you can use the formula $P(\bar{A}) = 1 - P(A)$, where $P(\bar{A})$ is the probability of event A not occurring, and $P(A)$ is the probability of event A occurring.

Addition Rule

The **addition rule** for probability is used for finding the probability of a compound event. Use the formula $P(A \cup B) = P(A) + P(B) - P(A \cap B)$, where $P(A \cap B)$ is the probability of both events occurring to find the probability of a compound event. The probability of both events occurring at the same time must be subtracted to eliminate any overlap in the first two probabilities.

Conditional Probability

Given two events A and B, the **conditional probability** $P(A|B)$ is the probability that event A will occur, given that event B has occurred. The conditional probability cannot be calculated simply from $P(A)$ and $P(B)$; these probabilities alone do not give sufficient information to determine the conditional probability. It can, however, be determined if you are also given the probability of the intersection of events A and B, $P(A \cap B)$, the probability that events A and B both occur. Specifically, $P(A|B) = \frac{P(A \cap B)}{P(B)}$. For instance, suppose you have a jar containing two red marbles and two blue marbles, and you draw two marbles at random. Consider event A being the event that the first marble drawn is red, and event B being the event that the second marble drawn is blue. If we want to find the probability that B occurs given that A occurred, $P(B|A)$, then we can compute it using the fact that $P(A)$ is $\frac{1}{2}$, and $P(A \cap B)$ is $\frac{1}{3}$. (The latter may not be obvious, but may be determined by finding the product of $\frac{1}{2}$ and $\frac{2}{3}$). Therefore $P(B|A) = \frac{P(A \cap B)}{P(A)} = \frac{1/3}{1/2} = \frac{2}{3}$.

Conditional Probability in Everyday Situations

Conditional probability often arises in everyday situations in, for example, estimating the risk or benefit of certain activities. The conditional probability of having a heart attack given that you exercise daily may be smaller than the overall probability of having a heart attack. The conditional probability of having lung cancer given that you are a smoker is larger than the overall probability of having lung cancer. Note that changing the order of the conditional probability changes the meaning: the conditional probability of having lung cancer given that you are a smoker is a very different thing from the probability of being a smoker given that you have lung cancer. In an extreme case, suppose that a certain rare disease is caused only by eating a certain food, but even then, it is unlikely. Then the conditional probability of having that disease given that you eat the dangerous food is nonzero but low, but the conditional probability of having eaten that food given that you have the disease is 100%!

Review Video: Conditional Probability
Visit mometrix.com/academy and enter code: 397924

Independence

The conditional probability $P(A|B)$ is the probability that event A will occur given that event B occurs. If the two events are independent, we do not expect that whether or not event B occurs should have any effect on whether or not event A occurs. In other words, we expect $P(A|B) = P(A)$.

This can be proven using the usual equations for conditional probability and the joint probability of independent events. The conditional probability $P(A|B) = \frac{P(A\cap B)}{P(B)}$. If A and B are independent, then $P(A \cap B) = P(A)P(B)$. So $P(A|B) = \frac{P(A)P(B)}{P(B)} = P(A)$. By similar reasoning, if A and B are independent then $P(B|A) = P(B)$.

Multiplication Rule

The **multiplication rule** can be used to find the probability of two independent events occurring using the formula $P(A \cap B) = P(A) \times P(B)$, where $P(A \cap B)$ is the probability of two independent events occurring, $P(A)$ is the probability of the first event occurring, and $P(B)$ is the probability of the second event occurring.

The multiplication rule can also be used to find the probability of two dependent events occurring using the formula $P(A \cap B) = P(A) \times P(B|A)$, where $P(A \cap B)$ is the probability of two dependent events occurring and $P(B|A)$ is the probability of the second event occurring after the first event has already occurred.

Use a **combination of the multiplication** rule and the rule of complements to find the probability that at least one outcome of the element will occur. This is given by the general formula $P(\text{at least one event occurring}) = 1 - P(\text{no outcomes occurring})$. For example, to find the probability that at least one even number will show when a pair of dice is rolled, find the probability that two odd numbers will be rolled (no even numbers) and subtract from one. You can always use a tree diagram or make a chart to list the possible outcomes when the sample space is small, such as in the dice-rolling example, but in most cases it will be much faster to use the multiplication and complement formulas.

Review Video: Multiplication Rule
Visit mometrix.com/academy and enter code: 782598

Union and Intersection of Two Sets of Outcomes

If A and B are each a set of elements or outcomes from an experiment, then the **union** (symbol ∪) of the two sets is the set of elements found in set A or set B. For example, if $A = \{2, 3, 4\}$ and $B = \{3, 4, 5\}$, $A \cup B = \{2, 3, 4, 5\}$. Note that the outcomes 3 and 4 appear only once in the union. For statistical events, the union is equivalent to "or"; $P(A \cup B)$ is the same thing as $P(A \text{ or } B)$. The **intersection** (symbol ∩) of two sets is the set of outcomes common to both sets. For the above sets

A and B, $A \cap B = \{3, 4\}$. For statistical events, the intersection is equivalent to "and"; $P(A \cap B)$ is the same thing as $P(A \text{ and } B)$. It is important to note that union and intersection operations commute. That is:

$$A \cup B = B \cup A \text{ and } A \cap B = B \cap A$$

PERMUTATIONS AND COMBINATIONS IN PROBABILITY

When trying to calculate the probability of an event using the $\frac{\text{desired outcomes}}{\text{total outcomes}}$ formula, you may frequently find that there are too many outcomes to individually count them. **Permutation** and **combination formulas** offer a shortcut to counting outcomes. A permutation is an arrangement of a specific number of a set of objects in a specific order. The number of **permutations** of r items given a set of n items can be calculated as ${}_nP_r = \frac{n!}{(n-r)!}$. Combinations are similar to permutations, except there are no restrictions regarding the order of the elements. While ABC is considered a different permutation than BCA, ABC and BCA are considered the same combination. The number of **combinations** of r items given a set of n items can be calculated as ${}_nC_r = \frac{n!}{r!(n-r)!}$ or ${}_nC_r = \frac{{}_nP_r}{r!}$.

Suppose you want to calculate how many different 5-card hands can be drawn from a deck of 52 cards. This is a combination since the order of the cards in a hand does not matter. There are 52 cards available, and 5 to be selected. Thus, the number of different hands is ${}_{52}C_5 = \frac{52!}{5! \times 47!} = 2{,}598{,}960$.

Review Video: Probability - Permutation and Combination
Visit mometrix.com/academy and enter code: 907664

TREE DIAGRAMS

For a simple sample space, possible outcomes may be determined by using a **tree diagram** or an organized chart. In either case, you can easily draw or list out the possible outcomes. For example, to determine all the possible ways three objects can be ordered, you can draw a tree diagram:

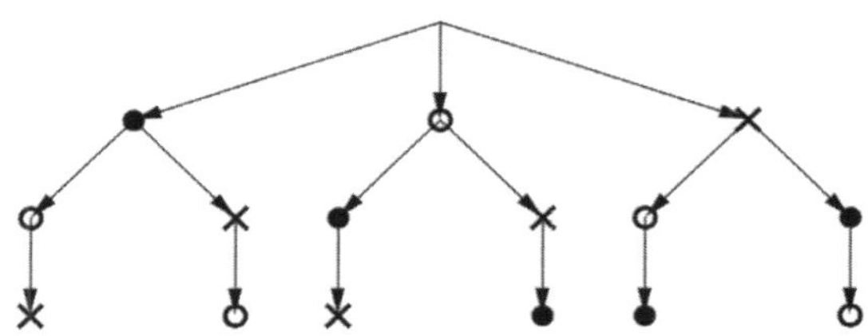

Review Video: Tree Diagrams
Visit mometrix.com/academy and enter code: 829158

You can also make a chart to list all the possibilities:

First object	Second object	Third object
●	x	o
●	o	x
o	●	x
o	x	●
x	●	o
x	o	●

Either way, you can easily see there are six possible ways the three objects can be ordered.

If two events have no outcomes in common, they are said to be **mutually exclusive**. For example, in a standard deck of 52 playing cards, the event of all card suits is mutually exclusive to the event of all card values. If two events have no bearing on each other so that one event occurring has no influence on the probability of another event occurring, the two events are said to be independent. For example, rolling a standard six-sided die multiple times does not change that probability that a particular number will be rolled from one roll to the next. If the outcome of one event does affect the probability of the second event, the two events are said to be dependent. For example, if cards are drawn from a deck, the probability of drawing an ace after an ace has been drawn is different than the probability of drawing an ace if no ace (or no other card, for that matter) has been drawn.

In probability, the **odds in favor of an event** are the number of times the event will occur compared to the number of times the event will not occur. To calculate the odds in favor of an event, use the formula $\frac{P(A)}{1-P(A)}$, where $P(A)$ is the probability that the event will occur. Many times, odds in favor is given as a ratio in the form $\frac{a}{b}$ or $a:b$, where a is the probability of the event occurring and b is the complement of the event, the probability of the event not occurring. If the odds in favor are given as 2:5, that means that you can expect the event to occur two times for every 5 times that it does not occur. In other words, the probability that the event will occur is $\frac{2}{2+5} = \frac{2}{7}$.

In probability, the **odds against an event** are the number of times the event will not occur compared to the number of times the event will occur. To calculate the odds against an event, use the formula $\frac{1-P(A)}{P(A)}$, where $P(A)$ is the probability that the event will occur. Many times, odds against is given as a ratio in the form $\frac{b}{a}$ or $b:a$, where b is the probability the event will not occur (the complement of the event) and a is the probability the event will occur. If the odds against an event are given as 3:1, that means that you can expect the event to not occur 3 times for every one time it does occur. In other words, 3 out of every 4 trials will fail.

Two-Way Frequency Tables

If we have a two-way frequency table, it is generally a straightforward matter to read off the probabilities of any two events A and B, as well as the joint probability of both events occurring, $P(A \cap B)$. We can then find the conditional probability $P(A|B)$ by calculating $P(A|B) = \frac{P(A \cap B)}{P(B)}$. We could also check whether or not events are independent by verifying whether $P(A)P(B) = P(A \cap B)$.

For example, a certain store's recent T-shirt sales:

	Small	Medium	Large	Total
Blue	25	40	35	100
White	27	25	22	74
Black	8	23	15	46
Total	60	88	72	220

Suppose we want to find the conditional probability that a customer buys a black shirt (event A), given that the shirt he buys is size small (event B). From the table, the probability $P(B)$ that a customer buys a small shirt is $\frac{60}{220} = \frac{3}{11}$. The probability $P(A \cap B)$ that he buys a small, black shirt is $\frac{8}{220} = \frac{2}{55}$. The conditional probability $P(A|B)$ that he buys a black shirt, given that he buys a small shirt, is therefore $P(A|B) = \frac{2/55}{3/11} = \frac{2}{15}$.

Similarly, if we want to check whether the event a customer buys a blue shirt, A, is independent of the event that a customer buys a medium shirt, B. From the table, $P(A) = \frac{100}{220} = \frac{5}{11}$ and $P(B) = \frac{88}{220} = \frac{4}{10}$. Also, $P(A \cap B) = \frac{40}{220} = \frac{2}{11}$. Since $\left(\frac{5}{11}\right)\left(\frac{4}{10}\right) = \frac{20}{110} = \frac{2}{11}$, $P(A)P(B) = P(A \cap B)$ and these two events are indeed independent.

Expected Value

Expected value is a method of determining the expected outcome in a random situation. It is a sum of the weighted probabilities of the possible outcomes. Multiply the probability of an event occurring by the weight assigned to that probability (such as the amount of money won or lost). A practical application of the expected value is to determine whether a game of chance is really fair. If the sum of the weighted probabilities is equal to zero, the game is generally considered fair because the player has a fair chance to at least break even. If the expected value is less than zero, then players are expected to lose more than they win. For example, a lottery drawing might allow the player to choose any three-digit number, 000–999. The probability of choosing the winning number is 1:1000. If it costs \$1 to play, and a winning number receives \$500, the expected value is $\left(-\$1 \times \frac{999}{1,000}\right) + \left(\$499 \times \frac{1}{1,000}\right) = -\$0.50$. You can expect to lose on average 50 cents for every dollar you spend.

Review Video: Expected Value
Visit mometrix.com/academy and enter code: 643554

Expected Value and Simulators

A die roll simulator will show the results of n rolls of a die. The result of each die roll may be recorded. For example, suppose a die is rolled 100 times. All results may be recorded. The numbers of 1s, 2s, 3s, 4s, 5s, and 6s, may be counted. The experimental probability of rolling each number will equal the ratio of the frequency of the rolled number to the total number of rolls. As the number of rolls increases, or approaches infinity, the experimental probability will approach the theoretical probability of $\frac{1}{6}$. Thus, the expected value for the roll of a die is shown to be $\left(1 \times \frac{1}{6}\right) + \left(2 \times \frac{1}{6}\right) + \left(3 \times \frac{1}{6}\right) + \left(4 \times \frac{1}{6}\right) + \left(5 \times \frac{1}{6}\right) + \left(6 \times \frac{1}{6}\right)$, or 3.5.

Permutations and Combinations

PERMUTATIONS AND COMBINATIONS

FACTORIALS

The **factorial** is a function that can be performed on any **non-negative integer**. It is represented by the ! sign written after the integer on which it is being performed. The factorial of an integer is the product of all positive integers less than or equal to the number. For example, 4! (read "4 factorial") is calculated as $4 \times 3 \times 2 \times 1 = 24$.

Since 0 is not itself a positive integer, nor does it have any positive integers less than it, 0! cannot be calculated using this method. Instead, 0! is defined by convention to equal 1. This makes sense if you consider the pattern of descending factorials:

$$5! = 120$$
$$4! = \frac{5!}{5} = \frac{120}{5} = 24$$
$$3! = \frac{4!}{4} = \frac{24}{4} = 6$$
$$2! = \frac{3!}{3} = \frac{6}{3} = 2$$
$$1! = \frac{2!}{2} = \frac{2}{2} = 1$$
$$0! = \frac{1!}{1} = \frac{1}{1} = 1$$

PERMUTATIONS

For any given set of data, the individual elements in the set may be arranged in different groups containing different numbers of elements arranged in different orders. For example, given the set of integers from one to three, inclusive, the elements of the set are 1, 2, and 3: written as $\{1, 2, 3\}$. They may be arranged as follows: 1, 2, 3, 12, 21, 13, 31, 23, 32, 123, 132, 213, 231, 312, and 321. These ordered sequences of elements from the given set of data are called **permutations**. It is important to note that in permutations, the order of the elements in the sequence is important. The sequence 123 is not the same as the sequence 213. Also, no element in the given set may be used more times as an element in a permutation than it appears as an element in the original set. For example, 223 is not a permutation in the above example because the number 2 only appears one time in the given set.

To find the number of permutations of r items from a set of n items, use the formula ${}_nP_r = \frac{n!}{(n-r)!}$. When using this formula, each element of r must be unique. Also, this assumes that different arrangements of the same set of elements yields different outcomes. For example, 123 is not the same as 321; order is important.

A special case arises while finding the number of possible permutations of n items from a set of n items. Because $n = r$, the equation for the number of permutations becomes simply $P = n!$ The same result is true for $r = n - 1$. Both of these cases are a result of the fact that 0! and 1! are both equal to 1.

If a set contains one or more groups of **indistinguishable or interchangeable elements** (e.g., the set $\{1, 2, 3, 3\}$, which has a group of two indistinguishable 3's), there is a different formula for finding distinct permutations of all n elements. Use the formula $P = \frac{n!}{m_1!m_2!...m_k!}$, where P is the number of permutations, n is the total number of elements in the set, and m_1 through m_k are the number of identical elements in each group (e.g., for the set $\{1, 1, 2, 2, 2, 3, 3\}$, $m_1 = 2$, $m_2 = 3$, and $m_3 = 2$). It is important to note that each repeated number is counted as its own element for the purpose of defining n (e.g., for the set $\{1, 1, 2, 2, 2, 3, 3\}$, $n = 7$, not 3).

To find the number of possible permutations of **any number of elements** in a set of unique elements, you must apply the permutation formulas multiple times. For example, to find the total number of possible permutations of the set $\{1, 2, 3\}$ first apply the permutation formula for situations where $n = r$ as follows: $P = n! = 3! = 6$. This gives the number of permutations of the three elements when all three elements are used. To find the number of permutations when only two of the three elements are used, use the formula ${}_nP_r = \frac{n!}{(n-r)!}$, where n is 3 and r is 2.

$${}_nP_r = \frac{n!}{(n-r)!} \Rightarrow {}_3P_2 = \frac{3!}{(3-2)!} = \frac{6}{1} = 6$$

To find the number of permutations when one element is used, use the formula ${}_nP_r = \frac{n!}{(n-r)!}$, where n is 3 and r is 1.

$${}_nP_r = \frac{n!}{(n-r)!} \Rightarrow {}_3P_1 = \frac{3!}{(3-1)!} = \frac{3!}{2!} = \frac{6}{2} = 3$$

Find the sum of the three formulas: $6 + 6 + 3 = 15$ total possible permutations.

Alternatively, the general formula for total possible permutations can be written as follows:

$$P_T = \sum_{i=1}^{n} \frac{n!}{(i-1)!}$$

Combinations

Combinations are essentially defined as permutations where the order in which the elements appear does not matter. Going back to the earlier example of the set $\{1, 2, 3\}$, the possible combinations that can be made from that set are 1, 2, 3, 12, 13, 23, and 123.

In a set containing n elements, the number of combinations of r items from the set can be found using the formula ${}_nC_r = \frac{n!}{r!(n-r)!}$. Notice the similarity to the formula for permutations. In effect, you are dividing the number of permutations by $r!$ to get the number of combinations, and the formula may be written ${}_nC_r = \frac{{}_nP_r}{r!}$. When finding the number of combinations, it is important to remember that the elements in the set must be unique (i.e., there must not be any duplicate items), and that no item may be used more than once in any given sequence.

Chapter Quiz

Ready to see how well you retained what you just read? Scan the QR code to go directly to the chapter quiz interface for this study guide. If you're using a computer, simply visit the online resources page at **mometrix.com/resources719/orelamgmath** and click the Chapter Quizzes link.

ORELA Practice Test

Want to take this practice test in an online interactive format?
Check out the online resources page, which includes interactive practice questions and much more: **mometrix.com/resources719/orelamgmath**

Multiple Choice

1. Quinn deposits $250 from each quarterly paycheck into a savings account. How much will he have deposited in the account after 5 years?

a. $4,800
b. $5,000
c. $5,200
d. $5,400

2. What is the value of x, shown in the diagram below?

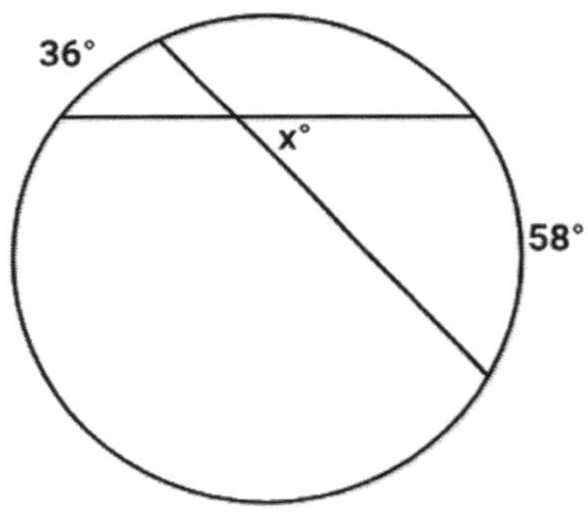

a. 11
b. 47
c. 51
d. 94

3. Amelia made an array of cubes that is 10 cubes by 10 cubes. Ms. Becker asked her to make the array 10 cubes by 15 cubes. Amelia picks up 15 more cubes. Which of the following statements best describes Amelia's error?

a. Amelia does not yet know number facts above 100
b. Amelia does not fully understand how changes to dimension affect measures of area and volume
c. Amelia does not fully understand the commutative property of multiplication
d. Amelia does not yet understand how to interpret mathematical models

4. The figure on the graph below represents a triangular walking path, beginning and ending at point x. If each unit represents 0.25 miles, which of the following best represents the total distance of the path?

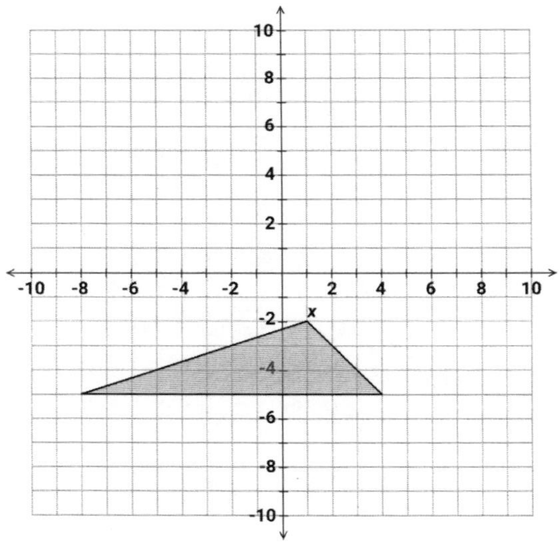

a. 6.4 miles
b. 12.2 miles
c. 15.8 miles
d. 25.7 miles

5. What is the interquartile range of the data below?

$$\{2, 4, 6, 8, 10, 12, 14, 16, 18, 20\}$$

a. 10
b. 11
c. 12
d. 13

6. In addition to math, in which other area of life do we tend to see patterns and relationships?

a. Works of art
b. Pieces of music
c. Articles of clothing
d. In all of these areas

7. Ms. Cho's students are solving multiplication and division word problems without using manipulatives. They read the following prompt:

"Daniel is feeding his neighbor's 4 dogs. Each dog gets 3 scoops of food daily. How many scoops of food will Daniel use each day?"

What question would Ms. Cho ask first in order to help her students express the word problem algebraically?

a. "How many scoops does Daniel need each day?"
b. "How can we calculate the total number of scoops?"
c. "Are we going to multiply or divide in this problem?"
d. "How many groups of 3 scoops will Daniel need?"

8. Jake rolls a standard six-sided die. What is the probability he rolls a number less than or equal to 5?

a. $\frac{5}{6}$
b. $\frac{1}{5}$
c. $\frac{3}{4}$
d. $\frac{1}{3}$

9. Which of the following statements is NOT true?

a. In a negatively skewed distribution, the tail is on the left.
b. In a positively skewed distribution, the median is greater than the mode.
c. In a normal distribution, the mean is at the peak.
d. In a positively skewed distribution, the median is closer to the third quartile than the first.

10. What is the converse of the statement below?

If it is Saturday, Joe buys a donut.

a. If it is not Saturday, Joe does not buy a donut.
b. If it is Saturday, Joe buys a donut.
c. If Joe buys a donut, it is Saturday.
d. If Joe does not buy a donut, it is not Saturday.

11. Which of the following accurately describes the set of integers?

a. The set of counting numbers
b. The set of counting numbers, plus zero
c. The set of numbers that may be written as the ratio of $\frac{a}{b}$, where $b \neq 0$
d. The set of counting numbers, zero, and the negations of the counting numbers

12. A car is accelerating. Which of the following accurately describes the appearance of the position-time graph?

a. It is a line with a positive slope.
b. It is a line with a negative slope.
c. It is a curve with an increasing slope.
d. It is a curve with a decreasing slope.

13. A convex three-dimensional figure has 9 edges and 6 vertices. How many faces does it have?

a. 4
b. 5
c. 6
d. 8

14. Ms. Dulik wants to introduce her students to the concept that a representation of data could be misleading. Which of the following pairs of data representation would best accomplish her goal? Select all that apply.

a. Two charts that contain different data sets
b. Two bar graphs that use different scales on the y-axis
c. Two drawings of a floor plan, with one marked "Not to scale"

15. Which of the following equations may be used to convert $0.\overline{4}$ to a fraction?

a. $10x - x = 4.\overline{4} - 0.\overline{4}$
b. $100x - x = 4.\overline{4} - 0.\overline{4}$
c. $10x - x = 44.\overline{4} - 4.\overline{4}$
d. $100x - 10x = 4.\overline{4} - 0.\overline{4}$

16. Which of the following measurements is the best approximation of 1,982 square centimeters?

a. 307 in^2
b. 865 in^2
c. 7,463 in^2
d. 12,787 in^2

17. What is the converse of the statement below?

If I go skiing, then it is winter.

a. If it is not winter, then I do not go skiing.
b. If it is winter, then I go skiing.
c. If it is not winter, then I go skiing.
d. If I go skiing, then it is not winter.

18. For which of the following data sets would the mean be an appropriate measure of center to use?

a. 7, 15, 20, 24, 27, 28, 31, 36, 41, 50
b. 6, 7, 8, 8, 9, 9, 10, 20, 34, 50
c. 5, 18, 30, 42, 43, 44, 46, 48, 49, 50
d. 8, 10, 12, 13, 14, 16, 20, 22, 24, 2200

19. Erika must choose her outfit from 4 pairs of jeans, 6 tops, and 5 pairs of shoes. How many possible outfits may she create with one item from each category?

a. 12
b. 24
c. 60
d. 120

20. Which of the following represents the net of a triangular prism?

a. 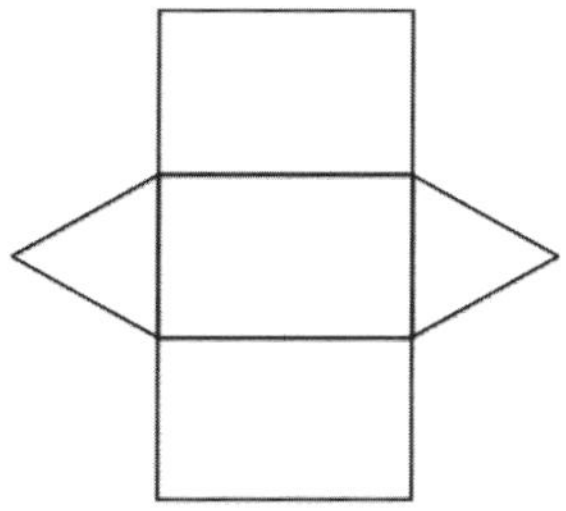

b. 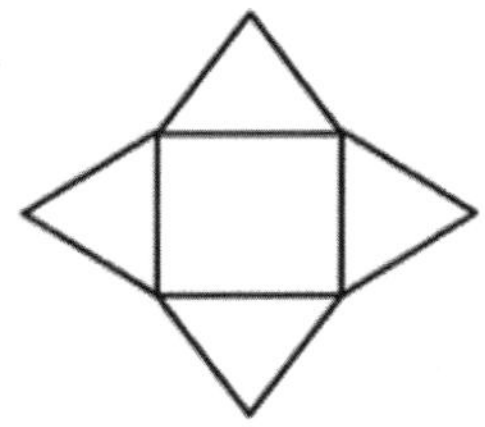

c. 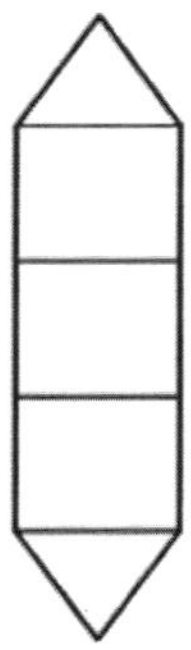

d. 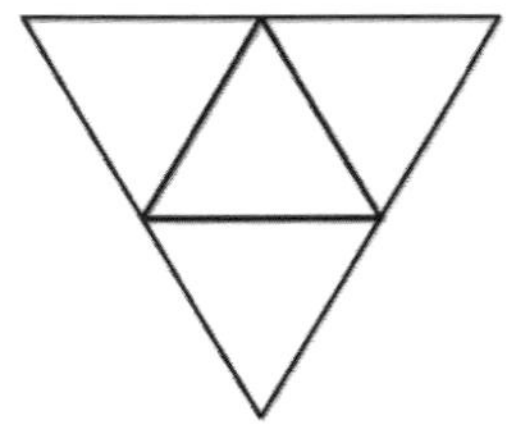

21. Which of these approaches to mathematics instruction is generally recognized as being the most effective?

a. Students need to memorize formulas and follow standard procedures for problem solving.
b. The use of mental math to solve complex problems should be discouraged.
c. Pencil-and-paper drills should be the primary focus of math class.
d. Students should be encouraged to solve problems in ways that make sense to them.

22. The table below shows the average amount of rainfall Houston receives during the summer and autumn months.

Month	Rainfall (in inches)
June	5.35
July	3.18
August	3.83
September	4.33
October	4.5
November	4.19

What percentage of rainfall in this timeframe occurs during October?

a. 13.5%
b. 15.1%
c. 16.9%
d. 17.7%

23. Which of the following graphs does NOT represent a function?

a. 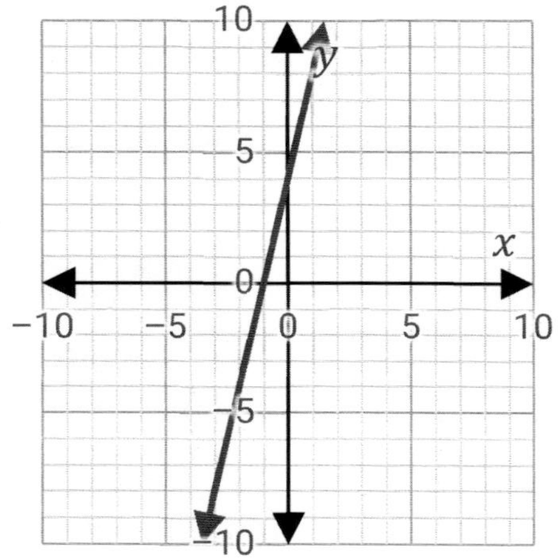

b. 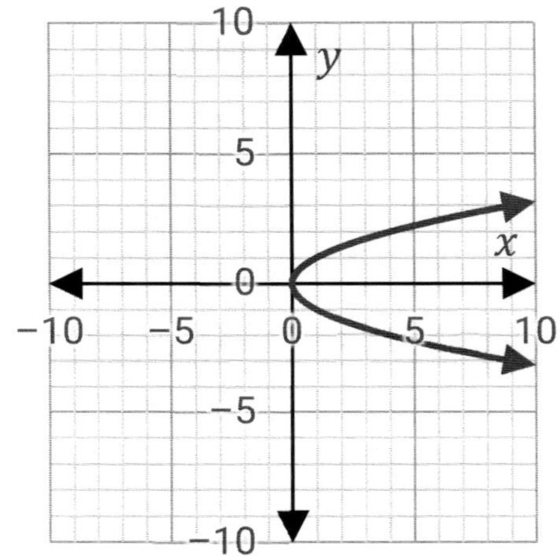

c. 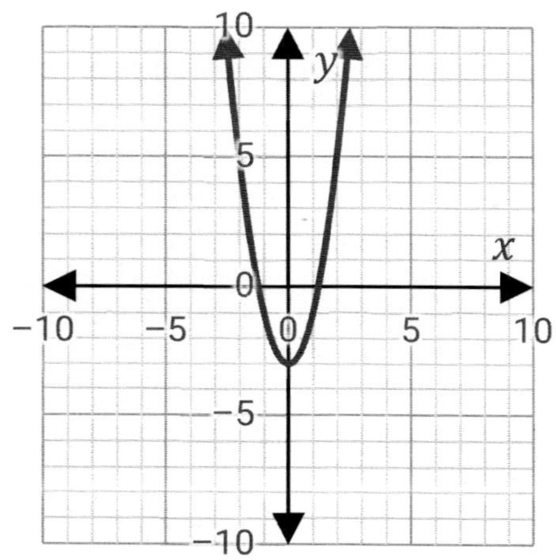

d. 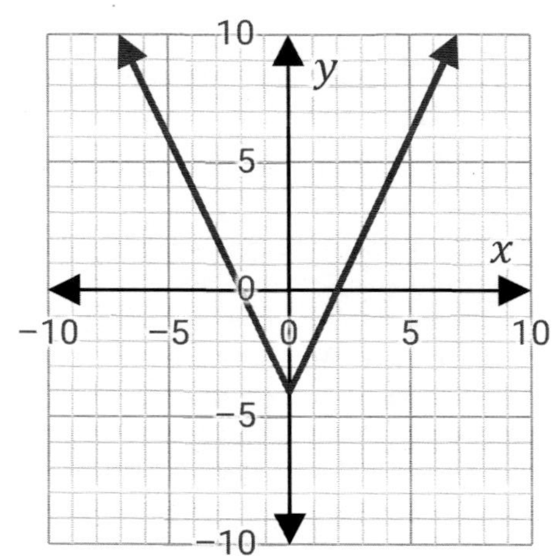

24. What is the area of the figure graphed below?

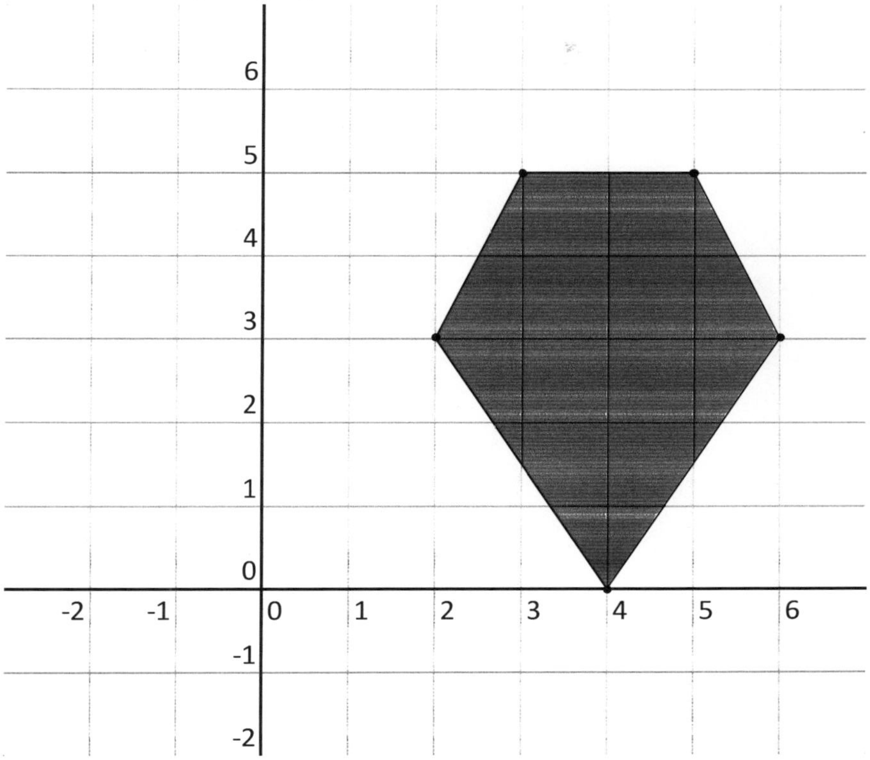

a. 11 units^2
b. 11.5 units^2
c. 12 units^2
d. 12.5 units^2

25. Tomas held out both hands, clearly showing ten digits – eight fingers and two thumbs – and proclaimed "I actually have nine fingers!" His friend Jerzy replied "No, you only have eight fingers!". To prove that he had nine fingers Tomas counted backward on one hand "8 – 7 – 6 – 5", then said "and 4 on this hand makes 9 fingers". This logic is incorrect because:

a. Tomas counted his thumb as one of the fingers
b Tomas was mixing ordinal and cardinal numbers
c. Tomas began counting from the left instead of the right
d. Tomas began counting from the right instead of the left

26. Which of the following represents a function?

a. $\{(2,1), (3,-2), (6,8), (-1,-4), (4,-1)\}$
b. $\{(8,5), (7,9), (3,1), (4,7), (3,6)\}$
c. $\{(-4,6), (2,3), (4,-2), (3,6), (-4,2)\}$
d. $\{(2,6), (6,1), (5,8), (2,-1), (-3,1)\}$

27. Eric's dietary plan consists of 4 different entrées, 5 different appetizers, and 3 different desserts. How many possible meals can he create that contain one entrée, one appetizer, and one dessert?

a. 12
b. 24
c. 60
d. 120

28. A bag contains 5 red marbles, 4 green marbles, and 3 yellow marbles. What is the probability that Fran pulls a red marble, keeps that marble in her possession, and then pulls a green marble?

a. $\frac{5}{36}$
b. $\frac{1}{6}$
c. $\frac{7}{132}$
d. $\frac{5}{33}$

29. On the first day of a new unit in math, a teacher gives students a pretest of the unit with 10 questions. All students score at least 80 percent on the pretest. If anything, what should the teacher do differently after seeing the pretest results?

a. Continue with the theme of the unit, but teach more advanced questions to further challenge students.
b. Move on to the next unit since students have mastered this one.
c. Reteach the questions that students missed on the pretest so that all students can score 100 percent.
d. Do not change instruction for the unit because the teacher is doing a good job already.

30. $g(x) = \frac{x}{x+3}$. What is the equation of the horizontal asymptote?

a. $y = 0$
b. $y = 0.5$
c. $y = 1$
d. $y = 3$

31. Marlon pays $45 for a jacket that has been marked down 25%. What was the original cost of the jacket?

a. $80
b. $75
c. $65
d. $60

32. Which type of function is represented by the table of values below?

x	y
−2	8
−1	2
0	0
1	2
2	8

a. Linear
b. Quadratic
c. Cubic
d. Exponential

33. On the first day of school, a teacher asks her students to create a list of tasks they feel they do well. What is the primary way in which this activity can help the teacher throughout the year?

a. It is a way for the teacher to informally assess her students' abilities and learn about their interests.
b. It serves as a tool the teacher can use to assess each student's background knowledge.
c. It will help the teacher know which students to pair in groups based on their skills.
d. It helps the students feel confident about what they already know.

34. A 4th-grade teacher introduces the math topic of decimals to the class. Together they preview their textbook chapter on decimals. The teacher assigns small groups of students to generate questions to which they want answers. Which of the following examples is the least helpful question to guide them at this stage of their learning?

a. "How are decimals important to us in our everyday lives?"
b. "How can decimals be whole numbers and also fractions?"
c. "In the number 467.912, which digit is in the hundredths?"
d. "Can we make fractions decimals, and decimals fractions?"

35. Alicia rolls a standard six-sided die. What is the probability she rolls a number greater than or equal to 3?

a. $\frac{5}{6}$
b. $\frac{3}{5}$
c. $\frac{3}{4}$
d. $\frac{2}{3}$

36. During February, the sixth-grade teachers at an elementary school decide to work together to teach a unit about the heart. The English teacher will have the students write essays about how the heart functions. The science teacher will teach the basic parts of the heart. The art teacher will have the students build models of the heart. The math teacher will have the students determine the amount of blood that flows through the heart in everyday units of capacity. Which of the following concepts are the teachers implementing?

a. Differentiated instruction
b. Formative assessment
c. Thematic units
d. Collaborative learning

37. Lilia needs to buy envelopes and stamps. Each box of envelopes costs \$2.50. Each book of stamps costs \$6.00. She intends to spend no more than \$30. Which of the following graphs represents the possible combinations of stamps and envelopes that she may purchase, if the y-axis represents books of stamps and the x-axis represents boxes of envelopes?

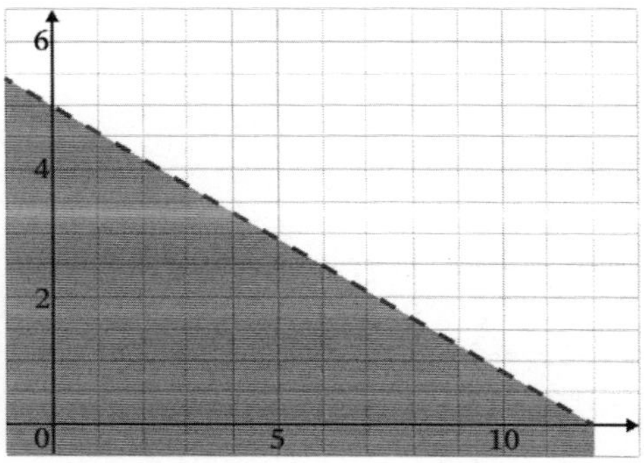

a.

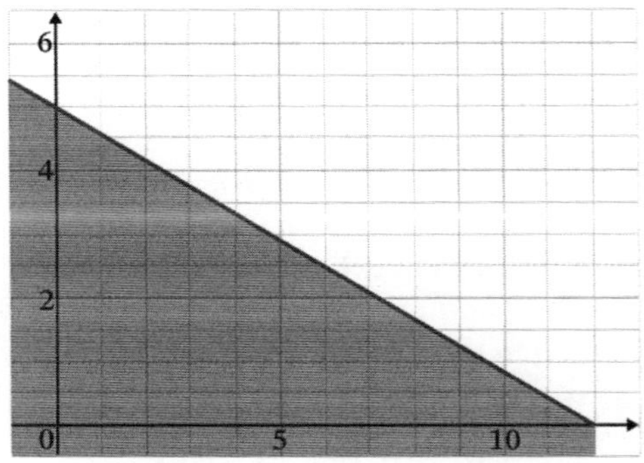

b.

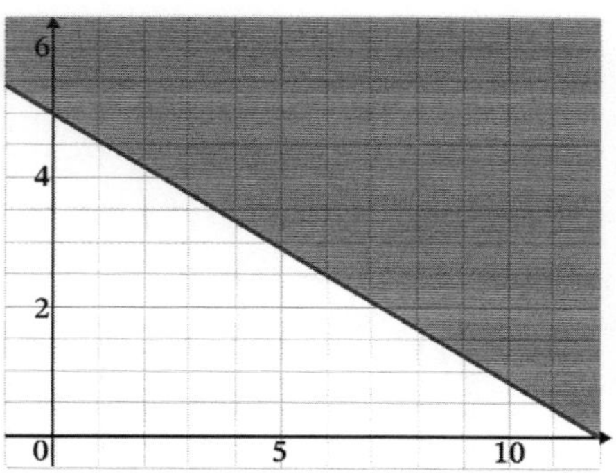

c.

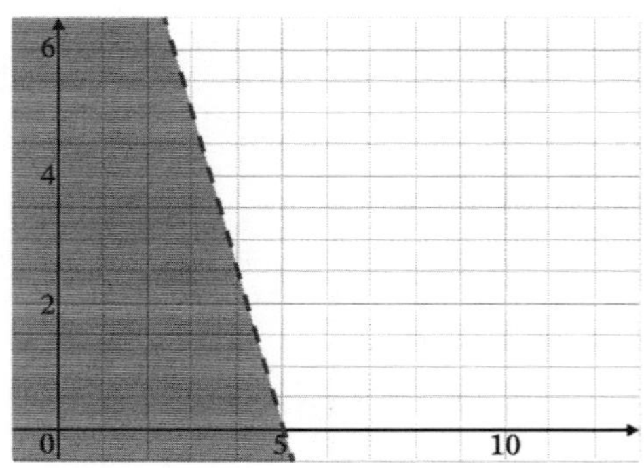

d.

38. The table below shows Marc's electricity bill during the summer and autumn months.

Month	Electricity bill
June	$143.79
July	$156.82
August	$160.91
September	$149.28
October	$127.56
November	$138.20

What percentage of Marc's total electricity costs were incurred during the month of October?

a. 13.5%
b. 14.6%
c. 15.9%
d. 17.2%

39. $g(x) = \frac{2x-3}{3-x}$**. What is the equation of the horizontal asymptote?**

a. $y = -2$
b. $y = -1$
c. $y = \frac{2}{3}$
d. $y = 3$

40. Kayla rolls a die and tosses a coin. What is the probability she gets an even number and heads?

a. $\frac{1}{6}$
b. $\frac{1}{4}$
c. $\frac{1}{3}$
d. 1

41. What is the interquartile range of the data below?

$$\{4, 7, 7, 9, 11, 14, 17, 18, 18, 23\}$$

a. 10
b. 11
c. 12
d. 13

42. Which of the following transformations has been applied to the triangle in Quadrant I?

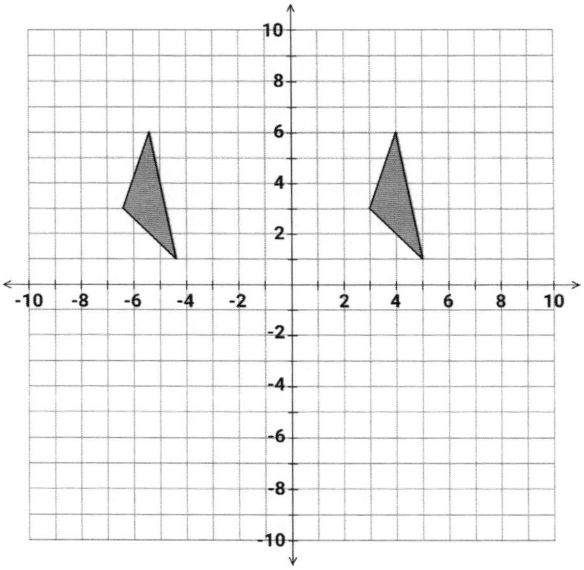

a. Translation
b. Rotation of 90 degrees
c. Reflection
d. Dilation

43. A city is at an elevation of 6,700 feet. Which of the following best represents the elevation in miles?

a. 0.77 miles
b. 1.27 miles
c. 1.56 miles
d. 1.89 miles

44. The table below shows the amount that Kelli spent on gas in the last six months.

Month	Gas expense
January	$78.85
February	$69.49
March	$97.48
April	$82.37
May	$72.86
June	$77.91

What percentage of the total gas costs for the six months was spent in March?

a. 17.5%
b. 18.1%
c. 19.9%
d. 20.4%

45. How many square inches of paper would it take to wrap a ball with a radius of 0.8 feet?

a. 8.04
b. 59.72
c. 437.96
d. 1,157.76

46. Which of the following formulas may be used to represent the sequence 9, 27, 81, 243, ...?

a. $y = 3x + 6$
b. $y = x + 9$
c. $y = 3^{x+1}$
d. $y = (x + 1)^3$

47. Mr. Abbas is teaching a lesson on calculating perimeter and area of two-dimensional shapes. He wants the lesson to engage his students in the practice of using appropriate tools strategically. Which of the following activities is best aligned with Mr. Abbas' objective?

a. Give each student a piece of yarn, a ruler, and a worksheet with pictures of various shapes. Ask students to use the tools to determine the perimeters and areas of the shapes.
b. Give each student a worksheet with pictures of various objects. Ask them to calculate the perimeter and area, showing their work.
c. Give each student a ruler or yardstick and a worksheet naming various objects in the classroom. Ask students to find the area of those objects.
d. Give each student a set of interlocking cubes and a worksheet naming various objects in the classroom. Ask students to estimate the area of those objects.

48. The two prisms shown below are similar. What is the measurement of x?

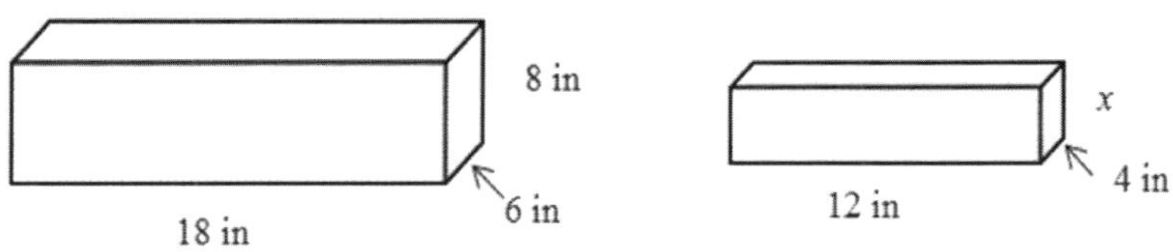

a. $4\frac{3}{4}$ in
b. $5\frac{1}{3}$ in
c. $5\frac{2}{3}$ in
d. $5\frac{3}{4}$ in

49. During a unit on integers, a math teacher administers a quiz to assess her students. A large proportion of students receive a failing grade. What is the best strategy the teacher should implement to correct this issue?

a. The teacher should individually conference with each student who failed to ask what they did to prepare for the quiz
b. The teacher should look at the quizzes to see what common mistakes were made, and work on reteaching the material to the students
c. The teacher should increase homework and encourage students to study at home
d. The teacher should ask students for input on what they still need to learn to be successful

50. What is the sum of the first 40 positive multiples of 3?

a. 615
b. 1,230
c. 2,460
d. 4,920

51. Jamal is placing an order for ink cartridges and reams of paper. Each ink cartridge costs \$7.50. Each ream of paper costs \$3. He intends to spend less than \$45. Which of the following graphs represents the possible combinations of paper and ink that he may purchase, if the x-axis represents ink cartridges and the y-axis represents reams of paper?

a.
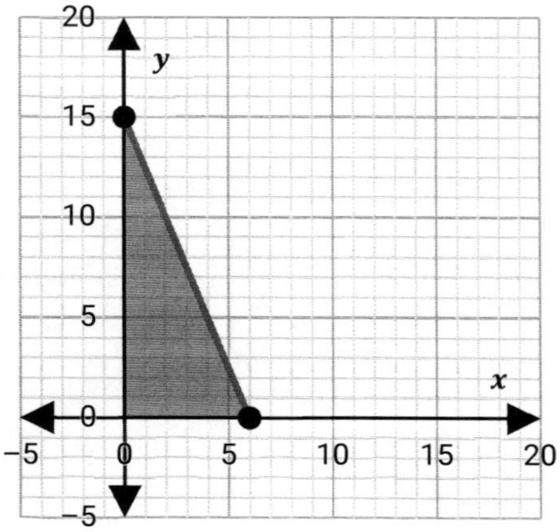

b.
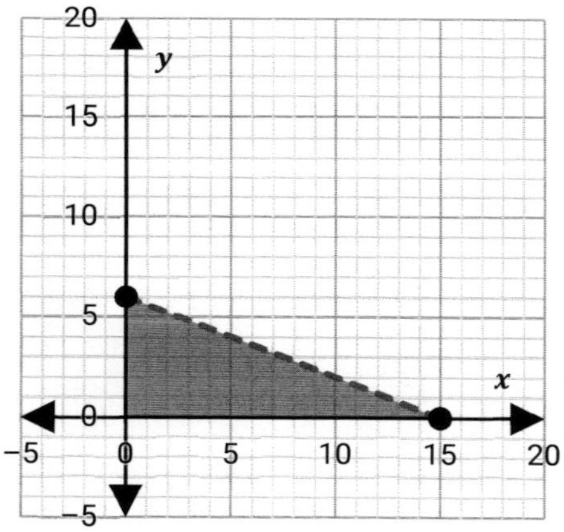

c.
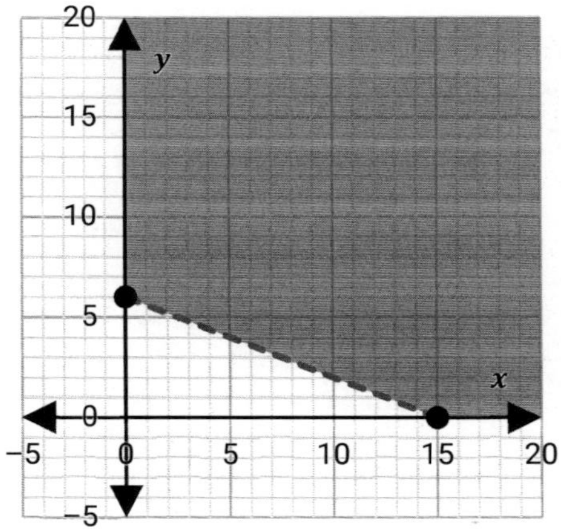

d.
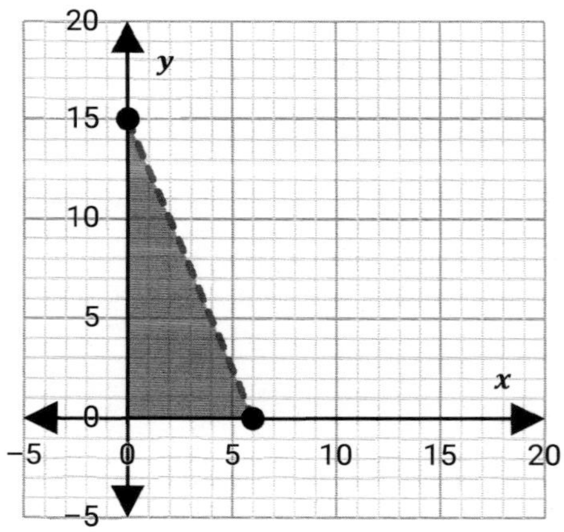

52. Which expression is represented by the diagram below?

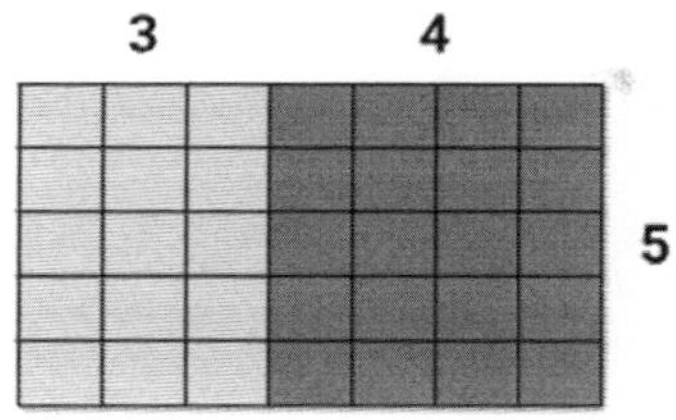

a. $5 + (3 + 4)$
b. $5 \times (3 \times 4)$
c. $5 + (3 \times 4)$
d. $5 \times (3 + 4)$

53. Which of the following functions has a limit of 0?

a. $f(x) = 6x^2 - 6$
b. $f(x) = \frac{4x}{x}$
c. $f(x) = \frac{3}{x+3}$
d. $f(x) = \frac{x+1}{x}$

54. A coin is tossed 300 times. How many of those tosses can you expect to show heads?

a. 50
b. 100
c. 150
d. 200

55. Josué is skip-counting by twos while looking at a page with the numbers 1–40 on a number line. His teacher overhears him say the following: "6, 8, 10, 12, 14, 16, 18, 20, 21, 22." Why might Josué skip-count accurately to a point and then begin to have trouble? Choose all that apply.

a. He is not able to use the concept of a number line.
b. He does not understand that the ones digits will follow the same patterns in numbers larger than 20 as they did in numbers 0–19.
c. He has good fluency counting by ones in numbers 0–20 but does not have the same fluency in numbers larger than 20.

56. Given the diagram below, which of the following theorems may be used to verify that lines *a* and *b* are parallel?

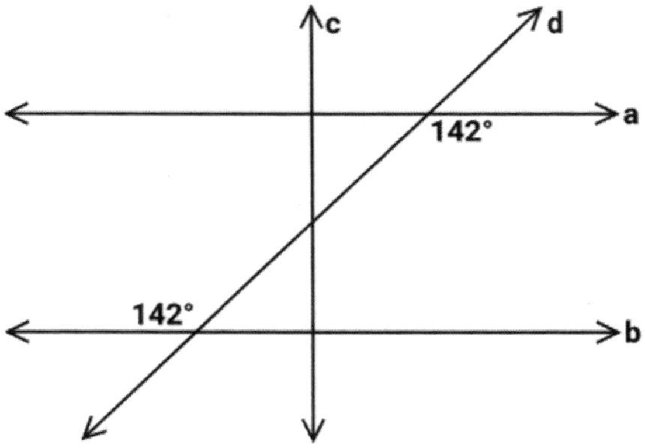

a. Alternate interior angles converse theorem
b. Alternate exterior angles converse theorem
c. Consecutive interior angles converse theorem
d. Corresponding angles converse theorem

57. Miss Wise asks her students to work in pairs to construct Venn diagrams which classify rational numbers, irrational numbers, real numbers, and integers. As she walks around the room to observe their progress, which of the following criterion could she use to assess the students' understanding of these sets of numbers?

a. The set of integers should include the other three sets.
b. The set of rational numbers should include the other three sets.
c. The set of irrational numbers should include the other three sets.
d. The set of real numbers should include the other three sets.

58. What is the approximate area of the shaded region in the figure shown below?

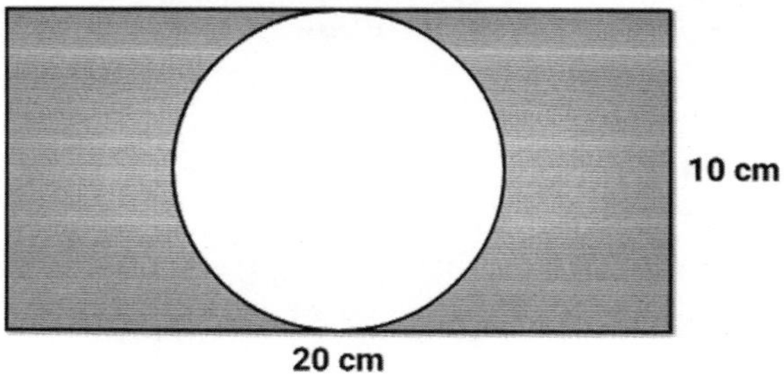

a. 122 cm^2
b. 144 cm^2
c. 176 cm^2
d. 193 cm^2

59. The initial term of a sequence is 7. Each term in the sequence is $\frac{8}{9}$ the amount of the previous term. What is the sum of the terms, as *n* approaches infinity?

a. 8
b. 28.25
c. 56
d. 63

60. An English language learner (ELL) student demonstrates a particular aptitude for mathematics in general, including arithmetic and algebra. The student's L1 is Arabic. The teacher incorporates vocabulary from the student's primary language into lessons in these subjects for the class and into assignments for the student. What is accurate about this practice?

a. It is appropriate for the student's content-area abilities but is not culturally or linguistically related.
b. It is culturally responsive and suits student content-area abilities but is not linguistically relevant.
c. It is culturally responsive, linguistically accessible, and meets the student's content-area abilities.
d. It is not culturally or linguistically meaningful and does not address student content-area abilities.

61. Which of the following accurately describes the set of rational numbers?

a. The set of whole numbers
b. The set of integers plus the roots of positive numbers
c. The set of numbers that may be written as the ratio of $\frac{a}{b}$, where $b \neq 0$
d. The set of counting numbers, zero, and the negations of the counting numbers

62. Given the two-way frequency table below, which of the following *best* represents *P*(bowl or chicken soup)?

	Chicken Soup	Vegetable Soup	Total
Cup	329	248	577
Bowl	642	591	1,233
Total	971	839	1,810

a. 70%
b. 75%
c. 80%
d. 85%

63. Amy saves $450 every 3 months. How much does she save after 3 years?

a. $4,800
b. $5,200
c. $5,400
d. $5,800

64. Which of the following correctly compares the sets of rational numbers and integers?

a. The sets of rational numbers and integers are equal.
b. The sets of rational numbers and integers are disjoint.
c. The set of integers is a subset of the set of rational numbers.
d. The set of rational numbers is a subset of the set of integers.

65. Of the following, which is an advantage of using analytical checklists for assessment?

a. They allow for differences in assessing individual tasks across students.
b. They require additional notes to identify other notable feats observed.
c. They save time and effort by listing tasks, skills, or behaviors in advance.
d. They sometimes constrain assessment to Yes/No without quality levels.

66. Which of the following tables contains points in an exponential function?

a.

x	y
−4	12
−1	−3
0	0
3	9
6	18

b.

x	y
−4	16
−1	2
0	1
2	$\frac{1}{4}$
5	$\frac{1}{32}$

c.

x	y
−3	9
−1	−3
0	0
2	6
5	15

d.

x	y
−4	−7
2	5
4	9
5	11
7	15

67. A seventh-grade math teacher considers how to use scaffolding to support her students in learning how to create a graph using a table. What is the first step she should take in scaffolding?

a. The teacher completes the task with student input.
b. The teacher models how to complete the task to students.
c. Both the teacher and student share the task duties.
d. Students try to complete the task on their own without support.

68. Which of the following best represents the measurement of x, shown in the right triangle below?

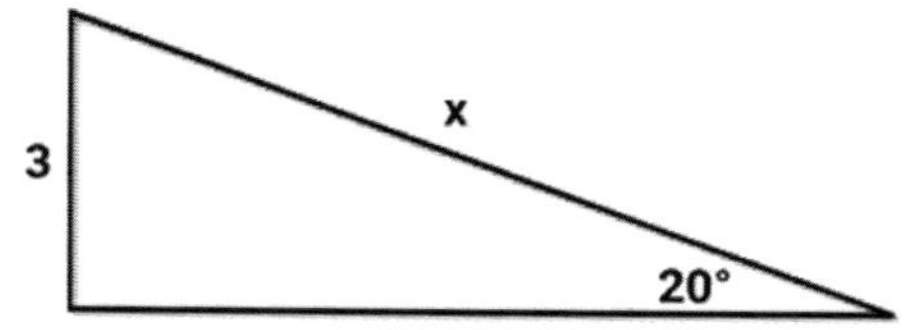

a. 8.8
b. 9.9
c. 10.1
d. 11.2

69. Which approach to assessment offers the strongest chance of quantifying what a student has learned?

a. Multiple choice tests
b. Essay questions
c. Combining several methods
d. Tests using a simple question-and-answer format

70. A missing object problem like the one below is one way of helping students learn what concept?

Truck car bike train truck car bike _______ truck car bike train

a. Counting
b. Problem solving
c. Manipulation of objects
d. Basic addition

71. A tree with a height of 15 feet casts a shadow that is 5 feet in length. A man standing at the base of the shadow formed by the tree is 6 feet tall. How long is the shadow cast by the man?

a. 1.5 feet
b. 2 feet
c. 2.5 feet
d. 3 feet

72. Jackson can decorate a cake in 3 hours. Eli can decorate the same cake in 2 hours. If they work together, how long will it take them to decorate the cake?

a. 0.8 hours
b. 1.2 hours
c. 1.5 hours
d. 1.8 hours

73. A math teacher spends a week teaching her advanced students a new concept. After giving a quiz, she notices that four students in her class are still having difficulty mastering the subject. What should this teacher's next step be in helping these students learn the material?

a. Assign the other students an enrichment activity while using targeted instruction with the group of four.
b. Allow the struggling students the opportunity to retake assessments to see if they improve.
c. Assign each of the four students a peer tutor in the classroom to assist them with their work.
d. Assign the students extra homework to see if their work improves in class.

74. Kendra uses the pie chart below to represent the allocation of her annual income. Her annual income is $40,000.

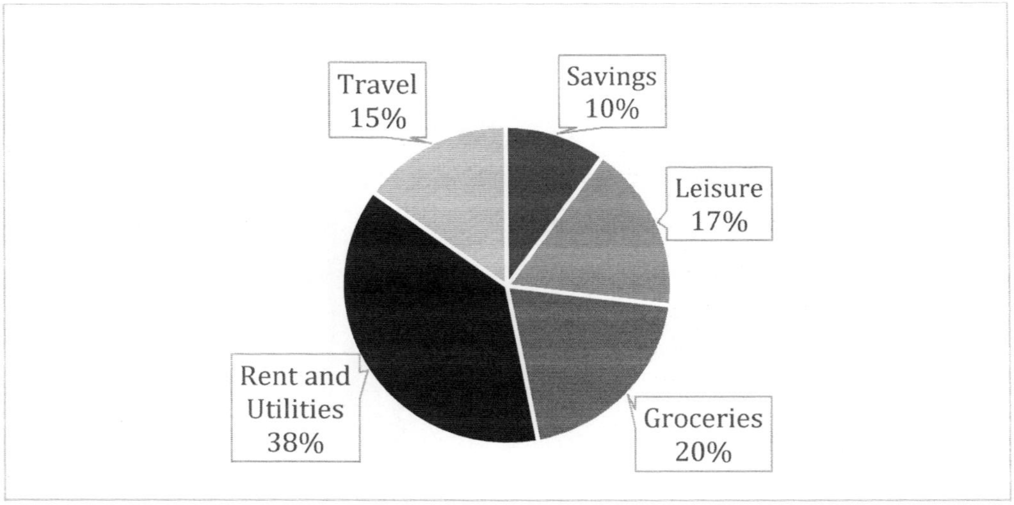

Which of the following statements is true?

a. The amount of money she spends on travel and savings is more than $11,000.
b. The amount of money she spends on rent and utilities is approximately $15,000.
c. The amount of money she spends on groceries and savings is more than $13,000.
d. The amount of money she spends on leisure is less than $5,000.

75. Which of the following statements is NOT true?

a. In a skewed distribution, the mean, median, and mode are all different.
b. In a positively skewed distribution, the mean will be to the right of the median.
c. In a positively skewed distribution, the mode is greater than the median.
d. The area under a normal curve is 1.

76. Each base of a triangular prism has a base length of 13 cm and a height of 15 cm. The height of the prism is 19 cm. What is the volume of the prism?

a. 1,572 cm^3
b. 1,684.75 cm^3
c. 1,775 cm^3
d. 1,852.5 cm^3

77. Given the diagram below, what is the measure of $\widehat{AC}$?

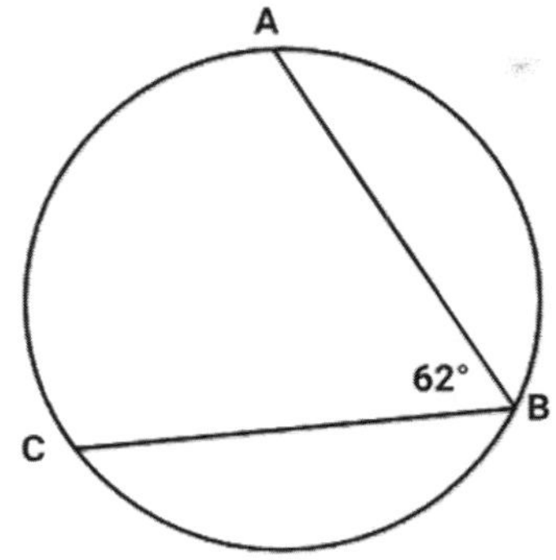

a. 31°
b. 62°
c. 124°
d. 208°

78. Mr. Fields wants to design an assessment method for his sixth grade math class that will help his students learn from their assignments and motivate them to improve. Which of the following approaches is most likely to accomplish this objective?

a Assigning ungraded homework problems and having weekly tests
b. Assigning nightly homework problems that are peer-graded in class the next day, and going over the answers after the assignment is graded
c. Assigning homework problems that are graded by the teacher and returned on the day before the unit test
d. Assigning nightly homework problems which are graded and returned with a correct answer key within three days

79. Eloisa's monthly car payment is $400. She is trading her car in for a newer model with a monthly payment of $525. What is the percent increase in her monthly payment?

a. 25%
b. 31.25%
c. 35%
d. 42.75%

80. Which of the following is an advantage of using anecdotal notes to assess student work?

a. Teachers can use these to collect information both during assessment and outside of it.
b. Unless a teacher adds norms or criteria for comparison, these notes lack standardization.
c. Teachers cannot record observations of individual student behaviors as with most tools.
d. Writing anecdotal notes can yield valuable insights but often lacks any supporting context.

81. A trough is in the shape of an inverted square pyramid. If the base has a side length of 9 feet and the height of the container is 2.5 feet, how many cubic feet of water can the trough hold?

a. 67.5 ft^3
b. 75 ft^3
c. 88.5 ft^3
d. 102 ft^3

82. Given that the two horizontal lines in the diagram below are parallel, which pair of angles is congruent?

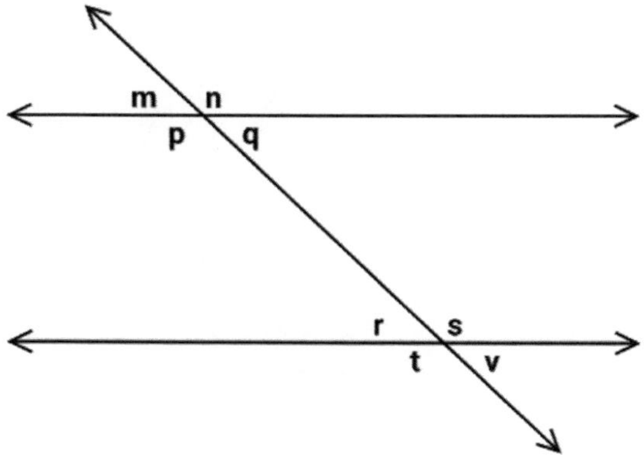

a. $\angle n$ and $\angle s$
b. $\angle m$ and $\angle p$
c. $\angle q$ and $\angle s$
d. $\angle s$ and $\angle v$

83. Eliza rolls a die. What is the probability she gets a 6 or an even number?

a. $\frac{1}{4}$
b. $\frac{1}{2}$
c. $\frac{2}{3}$
d. $\frac{3}{4}$

84. Which of the following graphs does NOT represent a function?

a. 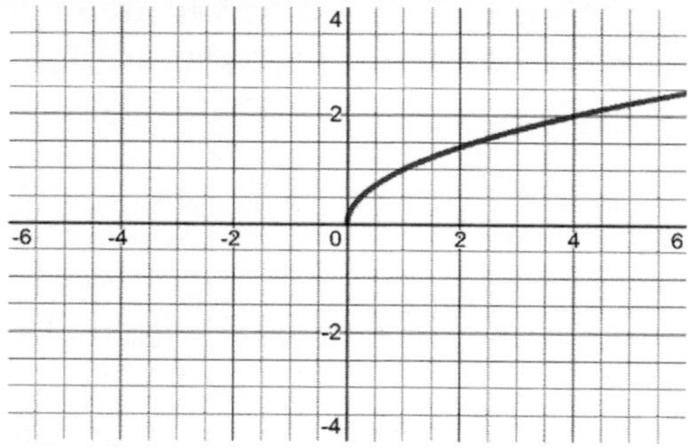

b. 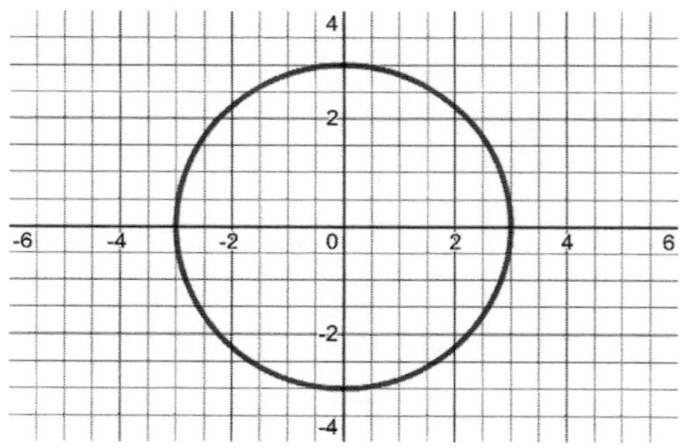

c. 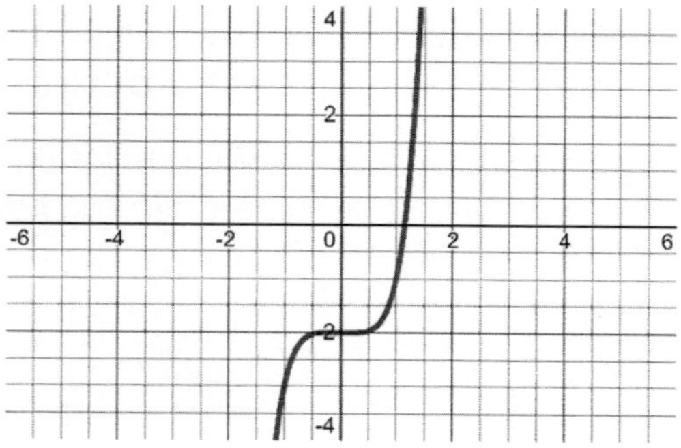

d. 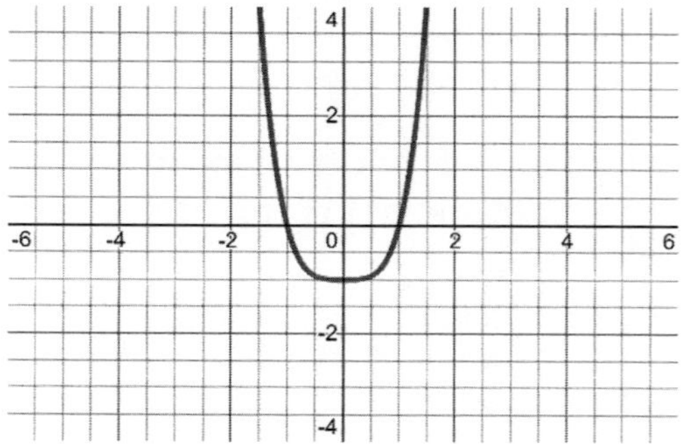

85. Zac eats $\frac{1}{9}$ of a bowl of chocolates. If the bowl contains $8\frac{4}{7}$ ounces of chocolate, how much chocolate does he eat?

a. $\frac{59}{60}$ ounces
b. $\frac{79}{80}$ ounces
c. $\frac{20}{21}$ ounces
d. $\frac{62}{63}$ ounces

86. A ball has a radius of 2.75 inches. Which of the following best represents the volume of the ball?

a. 79.9 in^3
b. 87.1 in^3
c. 96.5 in^3
d. 104.3 in^3

87. Lewis rolls a die and tosses a coin. What is the probability he gets a number less than 3 and heads?

a. $\frac{1}{6}$
b. $\frac{1}{4}$
c. $\frac{1}{3}$
d. 1

88. Which of the following best represents the value of x, shown in the triangle below?

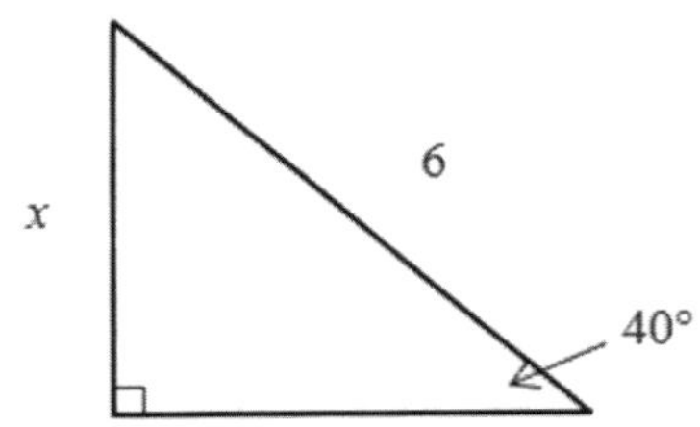

a. 2.6
b. 3.1
c. 3.9
d. 4.4

89. What is the midpoint of the line segment below?

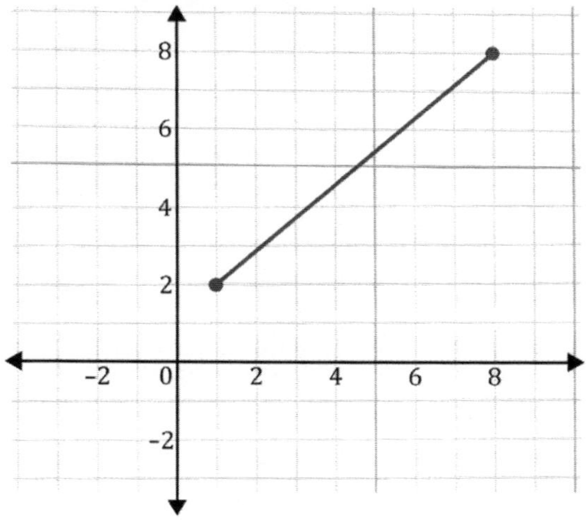

a. (3.5,4)
b. (4,4)
c. (4.5,5)
d. (5,5)

90. 225 students are surveyed. 95 of the students like only coffee. 30 of the students like only tea. 55 of the students like both coffee and tea. How many students like neither coffee nor tea?

a. 25
b. 35
c. 45
d. 55

91. Which of the following statements is true?

a. The set of rational numbers is a subset of the set of natural numbers.
b. The set of integers is a subset of the set of whole numbers.
c. The set of natural numbers is a subset of the set of whole numbers.
d. The set of complex numbers is a subset of natural numbers.

92. Which of the following statements is NOT true?

a. In most cases of a skewed distribution, the mean is pulled towards the longer tail.
b. In most cases of a skewed distribution, the mean is pulled towards the area with a higher frequency of scores.
c. In a normal distribution, the mean, median, and mode are the same value.
d. The area under a normal curve is 1.

93. The expression $2x^2 - 4x - 30$ is equal to the product of $(x - 5)$ and which other factor?

a. $(2x - 10)$
b. $(2x + 25)$
c. $(2x + 7)$
d. $(2x + 6)$

94. A third-grade student's continuous inability to do which of the following may indicate the potential need for further evaluation?

a. Multiply three-digit whole numbers
b. Compare and order fractions
c. Divide two-digit whole numbers
d. Add and subtract whole numbers within 100

95. What is $\lim_{x \to -4}\left(2x^3 - \frac{1}{2}x^2 - 1\right)$?

a. −137
b. −121
c. 4
d. 119

96. Mike is ordering a new polo shirt for each of his 12 employees, spending at least $150. Which of the following inequalities represents the possible costs?

a. $150 < 12p$
b. $150 > 12p$
c. $150 \leq 12p$
d. $150 \geq 12p$

97. Ms. King has taught rounding to her students by showing them to round to the nearest hundredth. She now wants her students to correctly identify situations in which this skill will be useful. She plans to present the students with two situations and ask them to choose the one that calls for rounding to the nearest hundredth. Which pair of situations would best match her objective?

a. Dividing a cookie equally between friends or buying enough cookies for 6 friends when the cookies come in packs of 4
b. Calculating sales tax on a pack of 4 cookies that costs $2.25 or dividing a cookie equally between friends
c. Calculating sales tax on a pack of 4 cookies that costs $2.25 or calculating the cost per cookie in a pack of 4 cookies that costs $2.25
d. Buying enough cookies for 6 friends when the cookies come in packs of 4 or calculating the cost of buying enough cookies for 6 friends when the cookies come in packs of 4 and cost $2.25

98. $g(x) = \frac{4-x}{x+6}$. What is the equation of the horizontal asymptote?

a. $y = 0$
b. $y = \frac{2}{3}$
c. $y = -1$
d. $y = -\frac{1}{6}$

99. Given the diagram below, what is the measure of the inscribed angle?

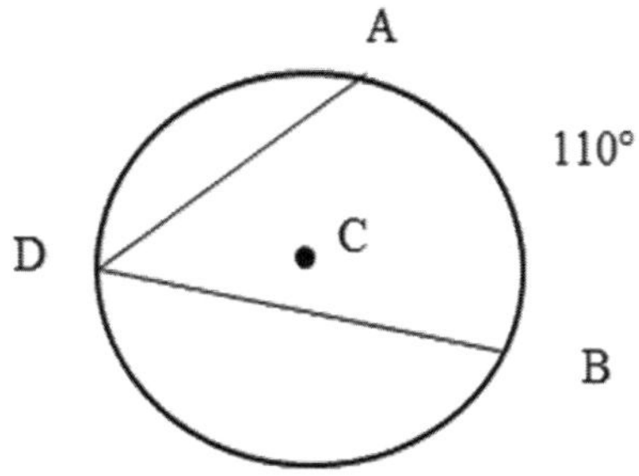

a. 37°
b. 45°
c. 55°
d. 57°

100. Raquel pays $12 per ticket to the fair, plus an $8 parking fee. Let x represent the number of tickets she buys. Which of the following equations represents her total cost y?

a. $y = 8x + 12$
b. $y = 20x$
c. $y = 12x + 8$
d. $y = 12x + 20$

Constructed Response

1. A family bought a new car for a purchase price of $32,000. The car will lose 15% of its value the day it is purchased, and the car will depreciate at a constant rate following that. The value of the car as a function of time can be modeled by $y = c - 0.09cx$, where y is the value of the car x years after the car was purchased and c is the value of the car after the initial 15% depreciation.

Part A: What is the value of the car 2 years after its purchase date? Show your work.

Part B: On an xy-grid, graph the value, y, of the car, as a function of x, where x represents the number of years after the purchase date, for $0 \leq x \leq 7$ years. Label the axes and show the scales used for the graph.

Part C: Use your graph to estimate the number of years, x, after the purchase date that the value of the car is $15,000. Label this point on your graph and indicate the approximate coordinates of the point.

Part D: Algebraically find the number of years, x, after the purchase date that the value of the car is exactly $15,000. Round your solution to the nearest tenth of a year. Show your work.

2. The diagram below shows the plan Berenice has for a triangular splash pad in a local city park. There will be three circles, each with a diameter of 6 feet, and the circles will be enclosed by an equilateral triangle with a side length of 30 feet. Berenice plans to have splash areas/fountains within the circles and walkways in the shaded areas of the triangle.

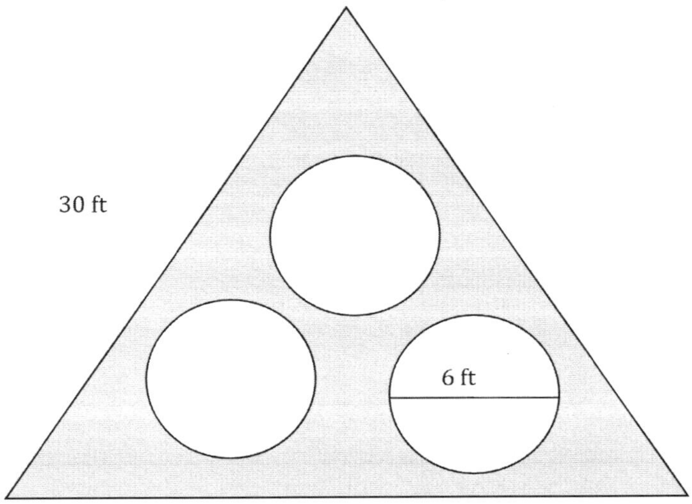

Part A: According to the diagram, what is the area, in square feet, available for Berenice to have for splash/fountain areas in the triangle? Show your work.

Part B: According to the diagram, what percentage of the triangle will be set aside for walkways? Show your work.

Part C: If a ball were to fall on a random point within the triangular splash pad, what is the probability that the ball would fall where Berenice plans to have splash areas? Show your work or explain your reasoning.

3. The heights of 19 students, in inches, are listed below.

48, 53, 53, 54, 55, 56, 59, 60, 60, 63, 63, 63, 64, 65, 66, 67, 69, 71, 73

Part A: For the numbers above, define and identify the median and the range.

Part B: Define and calculate the mean for the list of numbers above.

Part C: Draw a stem and leaf plot of the data using the tens digits as the stems and the units digits as the leaves.

Answer Key and Explanations

Multiple Choice

1. B: There are 60 months in 5 years. Since a quarter is 3 months, the following proportion may be written.

$$\frac{250}{3} = \frac{x}{60}$$

We solve by first cross-multiplying.

$$3x = 15{,}000$$

Then, divide both sides of the equation by 3.

$$x = 5{,}000$$

Therefore, Quinn will have deposited $5,000 into the account after 5 years.

2. B: The measure of an angle formed by intersecting chords inside a circle is equal to one-half of the sum of the measures of the intercepted arcs. Thus, $x = \frac{1}{2}(36° + 58°)$, or 47°.

3. B: Amelia underestimated how many more cubes she would need to fill an area when provided with a change in dimension. Choice C would also be correct if it referred to the distributive property rather than the commutative property of multiplication Amelia could have chosen the correct number of cubes if she understood that $10(15) = 10(10) + 10(5)$.

4. A: The perimeter of the triangle is equal to the sum of the side lengths. The length of the longer diagonal side may be represented as $d = \sqrt{(1 - -8)^2 + (-2 - -5)^2}$, which simplifies to $d = \sqrt{90}$. The length of the shorter diagonal side may be represented as $d = \sqrt{(4 - 1)^2 + (-2 - -5)^2}$, which simplifies to $d = \sqrt{18}$. The base length is 12 units. Thus, the perimeter is equal to $12 + \sqrt{90} + \sqrt{18}$, which is approximately 25.73 units. Since each unit represents 0.25 miles, the total distance of the path is equal to the product of 25.73 and 0.25, or approximately 6.4 miles.

5. A: The median of the lower half of the scores is 6. The median of the upper half of the scores is 16. The interquartile range is the difference between the first and third quartiles: $16 - 6 = 10$.

6. D: We tend not only to see or impose patterns and relationships where they do not objectively exist (like seeing animal, human, and object shapes in the clouds); we also deliberately create patterns and relationships in the works of art (A), pieces of music (B), and articles of clothing (C) we produce. We also notice and appreciate the patterns and relationships in these products that others have created.

7. B: The question in choice B prompts the students to begin thinking about what operations are needed to find the quantity the question is asking for. Choice A simply restates the question asked in the word problem. Choices C and D are clarifying questions that Ms. Cho could ask after students have begun thinking about how to calculate the total amount of food that Daniel needs.

8. A: The number of outcomes for the event is 5 (rolling a 1, 2, 3, 4, or 5), and the sample space is 6 (numbers 1–6). Thus, the probability may be written as $\frac{5}{6}$.

9. D: In a positively skewed distribution, the median is closer to the first quartile than the third.

10. C: If the statement is written in the form $p \rightarrow q$, then the converse is represented as $q \rightarrow p$. Thus, the converse should read, "If Joe buys a donut, it is Saturday."

11. D: The set of integers is represented as $\{\dots, -3, -2, -1, 0, 1, 2, 3, \dots\}$. The numbers 1, 2, 3, ... are counting numbers, or natural numbers. Thus, the set contains the counting numbers, zero, and the negations of the counting numbers.

12. C: The position of an accelerating car is changing according to a non-constant speed. Thus, the graph will show a curve with an increasing slope. The slope is increasing since it represents the velocity, and the velocity is increasing.

13. B: The relationship between number of faces, edges, and vertices is represented by Euler's Formula, $E = F + V - 2$. Substituting 9 for E and 6 for V gives: $9 = F + 6 - 2$, which simplifies to $9 = F + 4$. Thus, $F = 5$.

14. B, C: Both choice B and choice C require careful reading of scale and labels in order to properly interpret the visual data representation.

15. A: The repeating decimal may be converted to a fraction by writing:

$$\begin{array}{r} 10x = 4.\overline{4} \\ - \quad x = 0.\overline{4} \\ \hline \end{array}$$

which simplifies as $10x - x = 4.\overline{4} - 0.\overline{4}$.

16. A: Because 1 inch is 2.54 centimeters, the following proportion may be written and solved for x: $\frac{2.54^2}{1} = \frac{1,982}{x}$. So $(2.54)(2.54)x = 1,982$. Dividing both sides of the equation by 6.4516 gives $x \approx 307.21$. Thus, 1,982 square centimeters is approximately equal to 307 square inches.

17. B: If the statement is written in the form $p \rightarrow q$, then the converse is represented as $q \rightarrow p$. Thus, the converse should read, "If it is winter, then I go skiing."

18. A: Data sets B and C are asymmetrical: data set B is skewed toward lower values, and data set C is skewed toward higher values. This makes the mean a poor measure of center. Data set D is mostly symmetrical but has a large outlier. The mean is very sensitive to outliers and is not an appropriate measure of center for data sets that include them. Data set A is roughly symmetrical and has no outliers; the mean would be an appropriate measure of center here.

19. D: This is a counting problem. The possible number of outfits is equal to the product of the possibilities for each category. The product of 4, 6, and 5 is 120. Thus, there are 120 outfits that she may create.

20. A: The net of a triangular prism has three rectangular faces and two triangular faces, and the rectangular faces must all be able to connect to each other directly. This is shown in choice A.

21. D: Each student is a unique individual, and different students may have different ways of conceptualizing the same problem. In the long run, allowing students to find their own ways of thinking about mathematics will make them better problem solvers. Mental math has an important role to play in everyday life. Memorization and drills also have their place, but they should not be the central focus in the classroom.

22. D: Begin by calculating the total rainfall in the summer and autumn months by adding all the values together. The total rainfall is 25.38 inches. Then, create a ratio comparing the rainfall in October to the total rainfall. The ratio $\frac{4.5}{25.38}$ represents the fraction of rainfall received during October. Convert the ratio to a percentage by dividing the numerator by the denominator and then multiplying by 100: $\frac{4.5}{25.38} \approx 0.177 = 17.7\%$. Therefore, about 17.7% of the rain fell during October.

23. B: The vertical-line test can determine whether a graph is a function. If any vertical line passes through more than one point at once, then the graph is not a function. A vertical line will cross a sideways parabola at two points anywhere beyond the vertex, so choice B is not a function.

24. C: The area of a trapezoid may be calculated using the formula, $A = \frac{1}{2}(b_1 + b_2)h$. Thus, the area of the trapezoid is represented as $A = \frac{1}{2}(4 + 2)(2)$, which simplifies to $A = 6$. The area of the triangle is represented as $A = \frac{1}{2}(4)(3)$, which also simplifies to $A = 6$. Thus, the total area is 12 square units.

25. B: When mixing ordinal and cardinal numbers the point of reference shifts by one unit. Ordinal numbering begins from 1, while cardinal numbering begins from 0.

26. A: A function is a relation in which there is not any x-value that is mapped to more than one y-value. The only relation that represents a function is $\{(2,1), (3,-2), (6,8), (-1,-4), (4,-1)\}$.

27. C: This is a counting problem. The possible number of meals is equal to the product of the possibilities for each category. The product of 4, 5, and 3 is 60. Thus, there are 60 meals that he can create.

28. D: To find the probability of Fran pulling two different marbles from the bag, we must find the probability of each event occurring individually, then multiply those probabilities together. To start, we can determine that Fran's chance of pulling a red marble is $\frac{5}{12}$ because there are 5 red marbles out of a total 12 marbles. Next, we must determine that Fran has a $\frac{4}{11}$ chance of pulling a green marble because there are 4 green marbles out of a new total 11 marbles. Notice, our total must now be reduced to 11 because Fran has one marble already in her possession. When we multiply our two events, $\frac{5}{12} \times \frac{4}{11}$, we get $\frac{20}{132}$, or $\frac{5}{33}$. The chance of Fran pulling a red marble followed by a green marble is $\frac{5}{33}$.

29. A: When students show mastery at the beginning of a unit, teachers should still work to teach material on the same topic, but teach more advanced concepts at a higher level. This helps challenge students and keeps them engaged in the material, and it helps students become more proficient at the easier questions while also allowing them to work towards answering the more challenging questions. While the teacher could shorten the unit to move on to harder units, it probably is best that they continue teaching the unit so that students can reinforce their skills to build on their pre-existing knowledge for the next unit. It is important to allow students a chance to learn more advanced material when the planned material may be too simple for them, as it will help increase their confidence and allow students a chance to improve on what they know.

30. C: The horizontal asymptote is equal to the ratio of the coefficients of x, or $\frac{1}{1}$, which equals 1.

31. D: The original cost may be represented by the equation $45 = x - 0.25x$ or $45 = 0.75x$. Dividing both sides of the equation by 0.75 gives $x = 60$.

32. B: The points in the table represent a symmetrical curve mirrored across the y-axis. This is a parabola, or a quadratic equation.

33. A: Asking students to create lists of what skills they feel they are strong in allows the teacher to informally assess what students know as well as learn about the students' interests. This should allow the teacher to use their strengths and interests to better plan instruction and activities in the classroom. Asking students to create a list of strengths is not a formal assessment and should not be used as an all-inclusive tool to assess the background knowledge of a student. While the teacher may be able to use the list of tasks as a way of gauging what students know, self-reported strengths on the first day of school are insufficient to make grouping decisions. Although creating a list of tasks that students can perform may increase student confidence temporarily, it is not the ultimate purpose of this activity and is instead a secondary benefit.

34. C: Question A is good for meeting the standard of applying mathematics to real life. Question B is good for students to learn the conceptual relationship of decimals and fractions as different yet equivalent ways of expressing parts of a whole. Question D is good for students to learn how to convert fractions to decimals and vice versa. Question C is better asked by the teacher than by students—and not at this stage of their learning, but only after they understand place value.

35. D: The number of outcomes in the event is 4 (rolling a 3, 4, 5, or 6), and the sample space is 6 (numbers 1–6). Thus, the probability may be written as $\frac{4}{6}$, which simplifies to $\frac{2}{3}$.

36. C: Since all of the teachers are implementing a theme about the heart, this is a thematic unit. Collaborative learning typically involves students working in groups. Formative assessments are ways teachers assess learning while working on a unit. Differentiated instruction is the use of multiple methods when teaching a unit. Therefore, the correct answer is choice C.

37. B: The situation may be modeled by the inequality $2.5x + 6y \leq 30$. Isolating the y-term gives $6y \leq -2.5x + 30$. Solving for y gives $y \leq -\frac{5}{12}x + 5$. Thus, the line's y-intercept is 5, and the slope is $-\frac{5}{12}$. Since she wants to spend \$30 or less, the shading should be below the line. Because she can choose to spend exactly \$30 (i.e., ≤ 30 rather than < 30), the line should be solid, not dashed.

38. B: The total electricity cost is \$876.56. Thus, the ratio $\frac{127.56}{876.56}$ represents the percentage of electricity cost incurred during October. $\frac{127.56}{876.56} \approx 0.146$ or 14.6%.

39. A: The horizontal asymptote is equal to the ratio of the two coefficients of x, or $\frac{2}{-1}$, which equals –2.

40. B: The probability may be written as $P(E \text{ and } H) = P(E) \times P(H)$. The probability of rolling an even number on a die is $E = \frac{3}{6} = \frac{1}{2}$. The probability of a coin landing on heads is $H = \frac{1}{2}$. Substituting the probability of each event gives $P(E \text{ and } H) = \frac{1}{2} \times \frac{1}{2} = \frac{1}{4}$.

41. B: The median of the lower half of the scores (the first quartile) is 7. The median of the upper half of the scores (the third quartile) is 18. The interquartile range equals the difference between the first and third quartiles: $18 - 7 = 11$.

42. A: The triangle was translated across the y-axis. In translation, each point is moved the same distance and direction.

43. B: The following proportion may be written and solved for x: $\frac{5{,}280}{1} = \frac{6{,}700}{x}$. Thus, $x \approx 1.27$.

44. D: The total gas cost is $478.96. Thus, the ratio $\frac{97.48}{478.96}$ represents the percentage of gas cost spent in March. $\frac{97.48}{478.96} \approx 0.204$ or 20.4%.

45. D: The surface area of a sphere may be calculated using the formula $SA = 4\pi r^2$. Substituting 0.8 for r gives $SA = 4\pi(0.8)^2$, which simplifies to $SA \approx 8.04$. The surface area of the ball is approximately 8.04 square feet. There are twelve inches in a foot, so there are $12^2 = 144$ square inches in a square foot. In order to convert this measurement to square inches, then, the following proportion may be written and solved for x: $\frac{144}{1} = \frac{x}{8.04}$. This means $x \approx 1{,}157.76$. It would take approximately 1,157.76 square inches of wrapping paper to wrap the ball.

46. C: The ratio between successive terms is constant (3), so this is a geometric sequence. A geometric sequence is represented by an exponential function.

47. A: The teacher is looking to teach about perimeter and area, a skill which applies only to two-dimensional shapes. The activity the teacher wants to use focuses on "using appropriate tools strategically," which means that the students should have an assortment of tools to use.

A good activity for teaching those skills is to provide students with shapes to find the area of. Shapes can include straight-edged shapes like squares and triangles, but they can also include circles or other shapes with rounded edges. The activity that is chosen needs to accommodate a variety of shapes, which is why the students would need yarn to trace curved edges and a ruler to measure the yarn or straight edges. The teacher should also be interested in having all students do the same activity to help make grading fair and possible, so using worksheets is probably a good idea but not the only option.

Choices B, C, and D all refer to "objects," which are three-dimensional. Surface area and volume are skills that build on area and perimeter, so a teacher would want to be careful about selecting a task that can be easily followed by the students. Choice B focuses on calculation and not on using appropriate tools. Choice C provides only one tool for the job, and may not accommodate curved shapes. Choice D also only provides the students with one tool to use, and it is a tool that is not typically used in the real world. Choice A is the best option because it requires the students to think strategically about what tools to use to obtain a measurement and calculate both perimeter and area. Keep in mind that an activity does not need to target all objectives at once, so it is okay if the teacher decided to have the students calculate only perimeter or only area.

48. B: Since the figures are similar, the following proportion may be written and solved for x.

$$\frac{6}{4} = \frac{8}{x}$$
$$6x = 4 \times 8$$
$$6x = 32$$
$$x = \frac{32}{6} = 5\frac{2}{6} = 5\frac{1}{3}$$

Therefore, the measure of x is $5\frac{1}{3}$ in.

49. B: When a large portion of students receive a failing grade on an assessment, the teacher should take the time to look at common mistakes made on the assessment. The teacher then should work on reteaching the material to the students and administer another assessment to gauge student progress. If most students did not perform well on an assessment, instead of individually conferencing with students to see what they did to prepare, the teacher should first self-reflect to see what they did to get the students prepared for the assessment. While encouraging students to study at home is a good practice for improving student performance, increasing the amount of homework does not guarantee better performance. In fact, it may do the opposite if students' stress levels increase due to the increase in amount of work. Although asking students for input on what they need is something teachers should do, it is not the requirement of the student to know what they need to be successful within a classroom.

50. C: The value of the 40th term may be found using the formula $a_n = a_1 + (n-1)d$. Substituting the number of terms for n, the initial value of 3 for a, and the common difference of 3 for d gives: $a_{40} = 3 + (40-1)(3)$, which simplifies to $a_{40} = 120$. Now, the value of the 40th term may be substituted into the formula, $S_n = \frac{n(a_1+a_n)}{2}$, which gives: $S_{40} = \frac{40(3+120)}{2}$, which simplifies to $S_{40} = 2{,}460$.

51. D: The situation may be modeled by the inequality $7.5x + 3y < 45$. Isolating the y-term gives $3y < -7.5x + 45$. Solving for y gives $y < -2.5x + 15$. Thus, the y-intercept will be 15, the line will be dotted, and a test point of (0,0) indicates the shading should occur below the line.

52. D: The rectangular array represents the product of the side lengths of 5 and $(3+4)$.

53. C: As the denominator approaches infinity, the value of the function will get smaller and smaller and converge to 0.

54. C: The probability of tossing a coin and getting heads is $\frac{1}{2}$. Multiply the probability by the number of coin tosses: $\frac{1}{2} \times 300 = 150$. You would expect to get heads about 150 times.

55. B, C: Choices B and C each explain a possible reason that the student only begins to make mistakes after a certain number. If choice A were correct, we could expect him to make mistakes more consistently. It is common for a young student to have better fluency with and understanding of smaller numbers that they use more frequently and can easily represent visually or concretely.

56. A: The alternate interior angles have congruent angle measures, each measuring 142°. According to the alternate interior angles converse theorem, two lines are parallel if a transversal, intersecting the lines, forms congruent alternate interior angles.

57. D: Real numbers include all rational and irrational numbers. Rational numbers include all integers. The set of real numbers include the other three sets.

58. A: The area of the rectangle is equal to $10 \times (20)$, or 200 square centimeters. The area of the circle is equal to $\pi(5)^2$, or approximately 78.5 square centimeters. The area of the shaded region is equal to the difference of the area of the rectangle and the area of the circle, or $200 \text{ cm}^2 - 78.5 \text{ cm}^2$, which equals 121.5 cm^2. So the area of the shaded region is about 122 cm^2.

59. D: The sum of an infinite geometric series may be modeled by the formula $S = \frac{a}{1-r}$, where a represents the initial value and r represents the common ratio. Substituting the initial value of 7 and common ratio of $\frac{8}{9}$ into the formula gives $S = \frac{7}{1-\frac{8}{9}}$, which simplifies to $S = \frac{7}{\frac{1}{9}}$ or 63.

60. C: This practice is culturally responsive and linguistically accessible because the Arab culture significantly contributed to the development of arithmetic and algebra, and the Arabic language is the origin of many arithmetic concepts and algebraic terms. It is also appropriate to the student's content-area abilities because the student displays a particular aptitude for arithmetic and algebra. Incorporating Arabic into lessons as well as assignments can also inform other students about the origins of arithmetic, algebra, and their concepts and terminology. Therefore (A), (B), and (D) are all incorrect.

61. C: The set of rational numbers includes any number that can be written as a fraction of two integers, where the denominator is not zero.

62. D: Let $P(B)$ represent the probability of choosing a bowl, and let $P(CS)$ represent the probability of choosing chicken soup. Then $P(B \text{ or } CS) = P(B) + P(CS) - P(B \text{ and } CS)$. Substituting values from the table, we find: $P(B \text{ or } CS) = \frac{1{,}233}{1{,}810} + \frac{971}{1{,}810} - \frac{642}{1{,}810} = \frac{1{,}562}{1{,}810} = 0.8629834254 \ldots \approx 85\%$.

63. C: There are 36 months in 3 years. Since Amy saves \$450 every 3 months, the following proportion can be written to find the amount she saves in 36 months: $\frac{450}{3} = \frac{x}{36}$. Cross multiplying gives you the equation $3x = 16{,}200$, which can be solved for x. Dividing both sides of the equation by 3 gives $x = 5{,}400$.

64. C: The set of integers is represented as $\{\ldots -3, -2, -1, 0, 1, 2, 3, \ldots\}$. The set of rational numbers includes all of these as well as any number that can be represented as a fraction of two integers, where the denominator is not zero. So the set of integers is a subset of the set of rational numbers.

65. C: Using analytical checklists as assessment tools has the advantage of saving time and effort by listing the tasks, skills, or behaviors to be performed and assessed in advance. They also have the advantage of making assessment uniform across students relative to tasks and/or their components, making choice A incorrect. They do require additional notes for teachers to identify other notable feats that teachers observe students accomplishing beyond the designated tasks (B), but this is a disadvantage. Although some checklists include performance levels similarly to rubrics, others constrain assessment to Yes/No responses (D), another disadvantage.

66. B: The table represents part of a geometric sequence, with a common ratio of $\frac{1}{2}$, so it also represents points of an exponential function.

67. B: When considering scaffolding, teachers should first begin with modeling the task that they would like students to be able to accomplish independently. Students observe how the teacher completes the task. The next step is to then have the teacher complete the task with student input, with the teacher primarily completing the task. After that, teachers release responsibility to the students with guided support. Finally, the teacher allows the students to complete the task independently.

68. A: The following equation may be written and solved for x: $\frac{\sin 20°}{1} = \frac{3}{x}$. Cross-multiplying gives: $x = \frac{3}{\sin 20°}$, or $x \approx 8.8$.

69. C: Some students perform better on tests, while others do better with papers or other projects. Just as students learn in a variety of ways, their talents for expressing what they have learned also varies widely. A teacher that evaluates students based on a wide variety of tasks will come away with a much more complete picture of each student's strengths and weaknesses.

70. B: Problem solving. Because an item is missing in the middle of the pattern, the process asked for cannot be counting or addition. Solving this puzzle does not involve manipulation of objects.

71. B: The following proportion may be written and solved for x: $\frac{15\text{ ft}}{5\text{ ft}} = \frac{6\text{ ft}}{x\text{ ft}}$. Cross multiplying results in $15x = 30$. Dividing by 15 gives $x = 2$. Thus, the shadow cast by the man is 2 feet in length.

72. B: The situation may be modeled with the equation $\frac{1}{3} + \frac{1}{2} = \frac{1}{t}$, which simplifies to $\frac{5}{6} = \frac{1}{t}$. Thus, $t = \frac{6}{5} = 1.2$. If working together, it will take them 1.2 hours to decorate the cake.

73. A: When a small group of students have trouble understanding a new concept, a teacher should attempt to help remediate those students. One of the best ways to handle this during class time is to provide an enrichment activity to the students who have mastered the concept while providing additional instruction to the students still working on mastery. Retaking assessments can also be utilized, but material should be retaught before having students retake assessments. Taking the time to try to help students understand concepts is particularly important in math classrooms because concepts tend to build upon each other, and teachers should try to incorporate this into their lesson plans when students are struggling.

74. B: The amount she spends on rent and utilities is equal to 0.38($40,000), or $15,200, which is approximately $15,000.

75. C: The median is always greater than the mode in a positively skewed distribution.

76. D: The volume of a prism may be calculated using the formula $V = Bh$, where B represents the area of the base and h represents the height of the prism. The area of each triangular base is represented by $A = \frac{1}{2}(13)(15)$, so the area of each base is equal to 97.5 square centimeters. Substituting 97.5 for the area of the base and 19 for the height of the prism gives $V = (97.5)(19)$ or $V = 1{,}852.5$. The volume of the prism is 1,852.5 cm^3.

77. C: The measure of the inscribed angle is half of the measure of the intercepted arc. Since the inscribed angle measures 62°, the intercepted arc is equal to (62°)(2) or 124°.

78. B: Mr. Fields is most likely to achieve his objective by assigning nightly homework problems that are peer-graded in class the next day, and going over the answers after the assignment is graded. This method offers two advantages: first, students receive immediate feedback on their performance that they can begin to apply to future assignments and tests; second, students have the opportunity to find out why they arrived at wrong answers when the teacher goes over the problems. This makes assessment a positive experience that helps students perform better in the future, rather than making it a negative experience that de-motivates students by making them feel powerless to improve their performance.

79. B: The percent increase is represented as $\frac{525-400}{400}$, which equals 0.3125 or 31.25%.

80. A: One advantage of writing anecdotal notes as an assessment tool is that teachers can collect information about student skills and behaviors outside of assessment activities as well as during them; this can help teachers realize valuable insights into student learning and behavior. The fact that such notes are not standardized unless the teacher incorporates norms or criteria for comparison (B) is a disadvantage. An advantage of anecdotal notes is that teachers *can* record observations of individual student behaviors, which standard test instruments, checklists, and most other tools do *not* allow (C). The fact that anecdotal notes often lack any supporting context (D) is a disadvantage.

81. A: The volume of a pyramid may be calculated using the formula $V = \frac{1}{3}Bh$, where B represents the area of the base and h represents the height of the pyramid. Since the base is a square, the area of the base is equal to 9^2, or 81 square feet. Substituting 81 for B and 2.5 for h gives $V = \frac{1}{3}(81)(2.5)$, which simplifies to $V = 67.5$ ft³.

82. A: Since $\angle n$ and $\angle s$ are corresponding angles, they are congruent.

83. B: Since they are not mutually exclusive events, the probability may be written as $P(6 \text{ or } E) = P(6) + P(E) - P(6 \text{ and } E)$. Substituting the probability of each event gives $(6 \text{ } or \text{ } E) = \frac{1}{6} + \frac{1}{2} - \frac{1}{6}$, or $\frac{1}{2}$.

84. B: A vertical line will cross the graph at more than one point. Thus, it is not a function.

85. C: The amount Zac eats may be written as $\frac{1}{9} \times \frac{60}{7}$, which equals $\frac{60}{63}$ or $\frac{20}{21}$. Thus, he eats $\frac{20}{21}$ ounces of chocolate.

86. B: The volume of a sphere may be calculated using the formula $V = \frac{4}{3}\pi r^3$, where r represents the radius. Substituting 2.75 for r gives $V = \frac{4}{3}\pi(2.75)^3$, which simplifies to $V \approx 87.1$.

87. A: The probability may be written as $P(N \text{ and } H) = P(N) \times P(H)$. Substituting the probability of each event gives $(N \text{ and } H) = \frac{1}{3} \times \frac{1}{2}$, which simplifies to $\frac{1}{6}$.

88. C: The following equation may be written and solved for x: $\sin 40° = \frac{x}{6}$. Multiplying both sides of the equation by 6 gives: $6 \cdot \sin 40° = x$, or $x \approx 3.9$.

89. C: A line segment with endpoints (x_1, y_1) and (x_2, y_2) has the midpoint $\left(\frac{x_1+x_2}{2}, \frac{y_1+y_2}{2}\right)$. The line segment shown has endpoints (1,2) and (8,8), so its midpoint is $\left(\frac{1+8}{2}, \frac{2+8}{2}\right)$, or (4.5,5).

90. C: A Venn diagram may be drawn to assist in finding the answer.

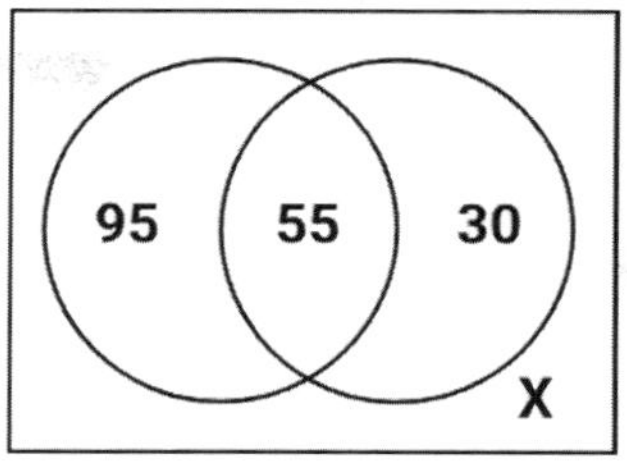

Since the set contains 225 total people, the solution is equal to $225 - (95 + 55 + 30)$, or 45 people

91. C: The set of natural numbers is contained within the set of whole numbers, and is hence a subset. Whole numbers include all positive integers plus zero, while natural numbers include all positive integers but not zero.

92. B: The mean is pulled towards the longer tail of a skewed distribution in most cases. It is not pulled towards the area with the higher frequency of scores. When outliers are added to an otherwise symmetric distribution, the mean shifts towards those outliers. For example, if a runner typically finishes a race in about 5 minutes and has one bad day in which she takes 20 minutes, that single outlier could significantly increase her mean time.

93. D: To factor an expression, find two expressions that multiply together to get the original expression. Since one factor is already known, we need to use the reverse FOIL method to find the other factor.

$$(x - 5)(?)$$

To find the first term of the second expression, divide $2x^2$ by x.

$$2x^2 \div x = 2x$$

$$(x - 5)(2x + ?)$$

To find the second term of the expression, we need to divide –30 by –5.

$$-30 \div -5 = 6$$

Therefore, the factored form of this expression is $(x - 5)(2x + 6)$. The FOIL method can be applied to multiply these two factors together and get the original expression, so the two factors are $(x - 5)$ and $(2x + 6)$.

94. D: While individual students will progress mathematically at varying rates, it is important that teachers are aware of the general expectations for typical skill development associated with each grade level. This is necessary for teachers to identify areas of individual student need so that they can differentiate and incorporate appropriate supports into instruction as well as recognize the signs of a developmental delay or learning disability. By third grade, students should have acquired a degree of fluency in adding and subtracting whole numbers within 100. If a student is continuously struggling to do so despite the implementation of academic supports, it may indicate a learning disability or delay and would require further evaluation to determine the need for special education services.

95. A: Evaluation of the expression for an x-value of −4 gives: $\left(2(-4)^3 - \frac{1}{2}(-4)^2 - 1\right)$, which equals −137.

96. C: Since he spends at least \$150, the relation of the number of polo shirts to the minimum cost may be written as $12p \geq 150$. Alternatively, the inequality may be written as $150 \leq 12p$.

97. B: This is the only choice that presents one real-world situation where students would apply rounding to the nearest hundredth (calculating sales tax). Choices A and D do not contain a real-world situation using rounding to the nearest hundredth. Both situations in choice C involve rounding to the nearest hundredth and therefore does not fit the teacher's objective for the activity.

98. C: The horizontal asymptote is equal to the ratio of the two coefficients of x, or $\frac{-1}{1}$, which equals −1.

99. C: The measure of the inscribed angle is half of the measure of the intercepted arc. Since the intercepted arc measures 110°, the inscribed angle is equal to $\frac{110°}{2}$ or 55°.

100. C: Raquel's total cost y equals \$12 times the number of tickets x, plus \$8. In other words, $y = 12x + 8$.

Constructed Response

1. Part A: First, according to the problem, c is the value of the vehicle after its initial depreciation. The vehicle's value drops by 15%, so:

$$c = \$32{,}000 - (0.15) \times (\$32{,}000) = \$27{,}200$$

Using this value for c and substituting it in, the value formula then becomes:

$$y = \$27{,}200 - (0.09) \times (\$27{,}200) \times (x)$$

Finally, solve for y when $x = 2$:

$$y = \$27{,}200 - (0.09) \times (\$27{,}200) \times (2)$$

$$y = \$27{,}200 - \$4896$$

$$\boldsymbol{y = \$22{,}304}$$

Part B: The graph of y will be linear since x is raised to the first power. Reordering the function to the $y = mx + b$ format, the y-intercept and slope are readily identifiable:

$$y = \$27{,}200 - \left(\frac{\$2{,}448}{year}\right) \times (x)$$

$$y = -\left(\frac{\$2{,}448}{year}\right) \times (x) + \$27{,}200$$

Thus, the slope is $-\$2448$ per year and the y-intercept is \$27,200. Plotting this function looks like the following.

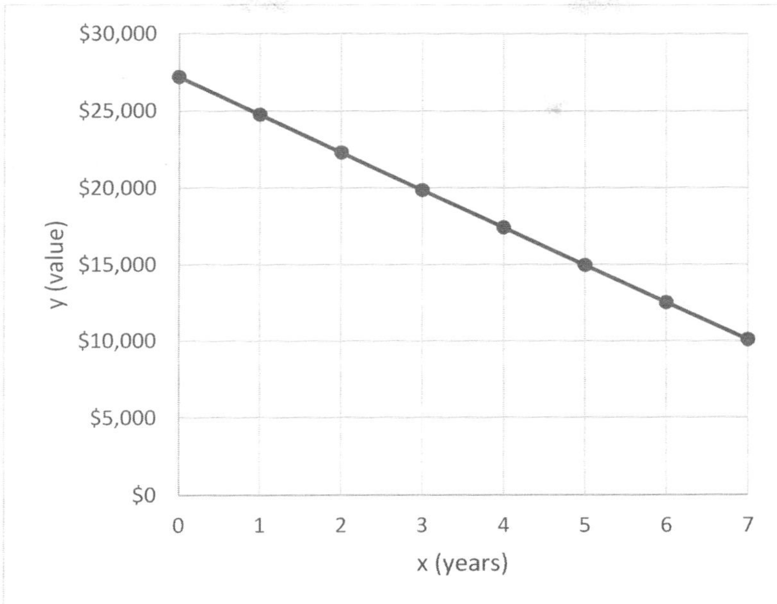

Part C: Using the axis values, it is readily apparent that the value function approaches \$15,000 when x is around **5 years**.

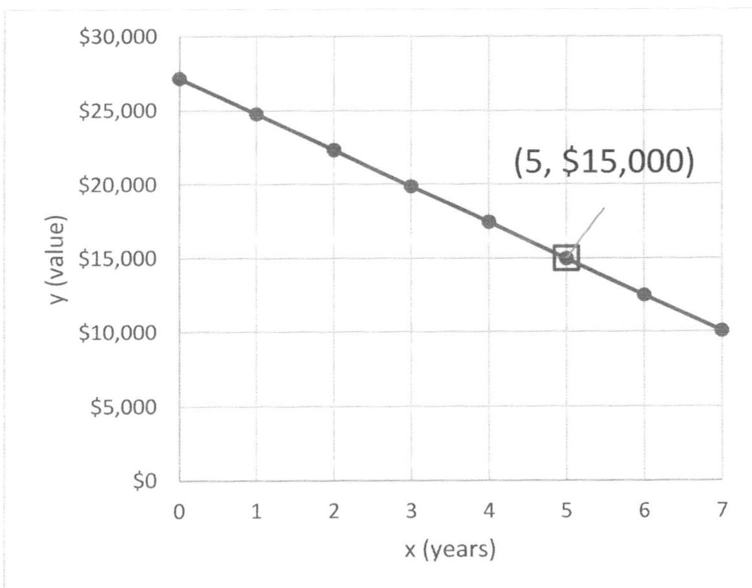

Part D: The way to use the function to find when the value is \$15,000 is to substitute it in for y and solve for x:

$$\$15{,}000 = \$27{,}200 - \left(\frac{\$2{,}448}{year}\right) \times (x)$$

$$-\$12{,}200 = -\left(\frac{\$2{,}448}{year}\right) \times (x)$$

$$\frac{-\$12{,}200}{-\left(\frac{\$2{,}448}{year}\right)} = x$$

$$x = 4.984\ years$$

After rounding to the tenths place: $\boldsymbol{x = 5.0\ years}$

2. Part A: The splash/fountain area is in the 3 circles only. To find the area, first find the radius of the circles: $r = \frac{d}{2} = \frac{6ft}{2} = 3ft$. Thus, the area for the splash/fountains is given by:

$$A_{splash} = 3 \times \pi \times r^2$$

Using $\pi \cong 3.14$:

$$A_{splash} \cong 3 \times 3.14 \times (3ft)^2$$

$$\boldsymbol{A_{splash} \cong 84.78ft^2}$$

Part B: The first step in finding the percent walkway area is to find the height of the triangle, since the area is known to be $\frac{1}{2} base \times height$. An equilateral triangle has equal side lengths and equal angles of 60°. Thus, we know that the base is 30ft and the height can be found by $h = 30ft \times \sin(60°)$.

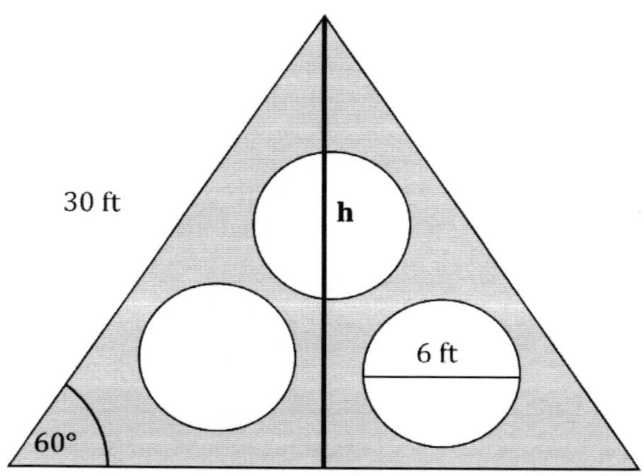

$h = 30ft \times \sin(60°) = 30ft \times \frac{\sqrt{3}}{2}$, so $h \cong 30ft \times \frac{1.73}{2}$. Thus, $height \cong 25.95ft$. Using this to find the area of the triangle minus the area of the circular splash/fountain sections will give us the total walking area:

$$A_{triangle} = \frac{1}{2} base \times height$$

$$A_{triangle} \cong \frac{1}{2}(30ft) \times (25.95ft) \cong 389.25ft^2$$

$$A_{walkway} = A_{triangle} - A_{splash}$$

$$A_{walkway} \cong (389.25ft^2) - (84.78ft^2)$$

$$A_{walkway} \cong 304.47ft^2$$

The percent walkway area is found by:

$$walkway\% = \frac{A_{walkway}}{A_{triangle}} \times 100\%$$

$$walkway\% \cong \frac{304.47ft^2}{389.25ft^2} \times 100\%$$

$$\boldsymbol{walkway\% \cong 78.2\%}$$

Part C: It can be assumed that a ball would have an equal chance of landing anywhere in the area of the triangle. Thus, the probability it will land in a splash/fountain area is given by:

$P_{splash} = \frac{A_{splash}}{A_{triangle}}$, so $\boldsymbol{P_{splash} \cong \frac{84.78ft^2}{389.25ft^2} \cong 0.218}$

3. Part A: The median of a list of values is the centermost value of an ordered list. If the ordered list has an even number of members, then the median is the average of the two centermost values. The range of a list is the difference of the highest value and the lowest value. Thus, for this problem, the **median is 63 inches and the range is 73 inches − 48 inches = 25 inches.**

Part B: The mean of a list of values is the sum of all the members of the list divided by the number of members in the list: mean =

$$\frac{48 + 53 + 53 + 54 + 55 + 56 + 59 + 60 + 60 + 63 + 63 + 63 + 64 + 65 + 66 + 67 + 69 + 71 + 73}{19}$$

Thus, **mean = 61.2 inches**

Part C: The stem and leaf plot would look like this:

4 | 8

Stem	Leaf
5	3 3 4 5 6 9
6	0 0 3 3 3 4 5 6 7 9
7	1 3

Online Resources

Due to our efforts to try to keep this book to a manageable length, we've created a link that will give you access to all of your online resources:

mometrix.com/resources719/orelamgmath

It's Your Moment, Let's Celebrate It!

Share your story @mometrixtestpreparation